The Liquid Metal
FAST BREEDER REACTOR

THE AEI
NATIONAL ENERGY PROJECT

The American Enterprise Institute's
National Energy Project was established in early 1974
to examine the broad array of issues
affecting U.S. energy demands and supplies.
The project will commission research into all important
ramifications of the energy problem—economic
and political, domestic and international, private
and public—and will present the results
in studies such as this one.
In addition it will sponsor symposia, debates, conferences,
and workshops, some of which will be televised.

The project is chaired by Melvin R. Laird,
former congressman, secretary of defense,
and domestic counsellor to the President,
and now senior counsellor of *Reader's Digest*.
An advisory council, representing a wide range of
energy-related viewpoints, has been appointed.
The project director is Professor Edward J. Mitchell
of the University of Michigan.

Views expressed are those of the authors
and do not necessarily reflect the views of
either the advisory council and others associated with
the project or of the advisory panels,
staff, officers, and trustees of AEI.

The Liquid Metal
FAST BREEDER REACTOR
An economic analysis

Brian G. Chow

American Enterprise Institute for Public Policy Research
Washington, D. C.

Brian G. Chow is associate professor of physics and observatory director at Saginaw Valley State College.

ISBN 0-8447-3192-7

National Energy Study 8, December 1975

Library of Congress Catalog Card No. 75-39899

Printed in the United States of America.

CONTENTS

1 **BACKGROUND FOR THE ASSESSMENT OF THE LIQUID METAL FAST BREEDER REACTOR PROGRAM**   **1**

The Administration and Funding of Energy Research and Development   1

Commercial Nuclear Reactors and the Liquid Metal Fast Breeder Reactor   6

The Liquid Metal Fast Breeder Reactor Program in the United States   10

Cost-Benefit Analyses by the Atomic Energy Commission   15

Summary of the Findings of this Study   18

2 **DISCOUNT RATE**   **21**

Discount Rates Used by the Atomic Energy Commission   21

Discount Rate Determination   22

3 **URANIUM SUPPLY SCHEDULES**   **29**

Uranium Resources in the United States   29

Correlation between Exploration Activities and Uranium Price   32

Underestimation of Present Uranium Resources   35

Uranium Supply-versus-Price Schedules for this Study   36

4 PLANT CAPACITIES, ELECTRICAL ENERGY
  DEMAND, AND PLANT MIX   39

   Overrestriction of High-Temperature Gas Reactor
     Plant Capacity   39
   Plant-Capacity Constraints Used in the
     Present Study   41
   Projections of Electrical Energy Demand   43
   The Role of Coal-Fired Plants in the Future
     Electric Power System   46
   Nuclear Power Forecasts by the Atomic Energy
     Commission   48

5 COST PROJECTIONS BY THE ATOMIC ENERGY
  COMMISSION   51

   Fuel-Cycle Costs of Light-Water Reactors   51
   Reactor Characteristics of Nuclear Reactors   54
   Plant Capital-Cost Differential   57
   Costs of the Liquid Metal Fast Breeder
     Reactor Program   62
   The AEC's Tradition of Cost Underestimation   65

6 CONCLUDING REMARKS   69

   The Stauffer-Palmer-Wyckoff Report on
     Fast Breeder   69
   Findings   71

APPENDIX: Present Worth of the LMFBR's Past Costs   75

1

BACKGROUND FOR THE ASSESSMENT OF THE LIQUID METAL FAST BREEDER REACTOR PROGRAM

The Administration and Funding of Energy Research and Development

Until recently the American people have been blessed with abundant supplies of fairly inexpensive energy, supplies which have been used in a wide variety of ways to sustain our high standard of living. With only 6 percent of the world's population, we consume approximately one-third of the world's annual energy supply. Then came the Arab oil embargo of October 1973. The general public suddenly recognized the economic consequences of high-cost energy and began to wonder about the causes of the energy crisis. Did the crisis result from dwindling energy resources or from an ill-coordinated national energy research and development program? Edward J. Mitchell sums it up as follows:

> . . . through the 1950s, such government-enforced mechanisms as market-demand prorationing and import quotas resulted in artificially high U.S. energy prices, huge domestic petroleum surpluses, heavy consumer costs, and production inefficiencies. Then, in the 1960s and '70s, the situation shifted from surplus to shortage with the gradual imposition of price controls. Energy prices were held far below competitive market level, a policy that stimulated consumption, dried up supplies, and eventually led to the current crisis.[1]

I wish to thank Dr. Edward J. Mitchell, director of AEI's National Energy Project, for his valuable suggestions, support and encouragement, the American Enterprise Institute itself for the research grant that helped support this study, and the staff of the University of Michigan's Graduate School of Business Administration for hospitality during my stay. My appreciation also goes to Mr. Wallace Hobkirk for his comments and Mrs. Hobkirk for her assistance. Finally, I am especially grateful to my wife, Pauline, and our daughter, Kira, for their patience and understanding.

[1] Edward J. Mitchell, *U.S. Energy Policy: A Primer* (Washington, D. C.: American Enterprise Institute, 1974), abstract.

Whether one agrees with such an assessment is irrelevant here. It is, however, important to realize that the government-enforced national energy policy affects our future and our children's future in a most significant way. Energy is a basic commodity fueling our economy, and no one can escape its influence. The sharply rising price of imported oil, from $2.65 per barrel before the embargo dictated by the thirteen-member Organization of Petroleum Exporting Countries (OPEC) to $10.80 per barrel currently, has already resulted in higher costs for practically every product and service.

The government is in the midst of launching multi-billion-dollar programs to meet the short- and long-term demand of energy. In light of the nationwide concern, these programs are now being recommended to and favorably reviewed by Congress, industry, and the general public. Programs aimed at resolving short-term energy problems may have to be enacted expeditiously. But proposals for the creation or acceleration of long-term programs should not be endorsed until the suggestions and attitudes of individuals and organizations with different perspectives can be thoroughly considered. This is especially true for programs that would require a substantial amount of the total energy budget and whose cost and benefit projections and extrapolations rely on calculations that reach far into the future. Once established, it would be difficult to change their courses. Therefore, the economic, social, and environmental impacts of these programs should first be thoroughly investigated and contrasted with those of their alternatives.

In the past, energy research and development (ER & D) have been administered by different agencies, such as the Atomic Energy Commission (AEC), the Office of Coal Research of the Department of Interior, "energy centers" in the Bureau of Mines, and the National Science Foundation's research and development on solar and geothermal energy. For twenty years we have been in need of a coordinated national ER & D program. The recently established Energy Research and Development Administration (ERDA), which will eventually become the Department of Energy and Natural Resources and absorb the Department of Interior, is a major step forward. In the interim, still another council, the Energy Resources Council (ERC), headed by Secretary of the Interior Rogers C. B. Morton, is authorized under the ERDA legislation to aid in the transition of agencies and to be responsible for the coordination and supervision of national energy policy.

ERDA's first administrator is Robert C. Seamans, Jr. There are six assistant administrators responsible, respectively, for fossil-energy development, nuclear-energy development, environmental safety, solar-energy research, geothermal and advanced energy systems, and national security and conservation. In addition to ERDA, the Nuclear Regulatory Commission (NRC) was created. Headed by William A. Anders, it will have three major parts: the Office of Nuclear Reactor Regulation, the Office of Nuclear Safety and Safeguards, and the Office of Nuclear Regulatory Research.

The regulatory and licensing functions of the defunct AEC were taken over by NRC, while its other functions were retained by ERDA. It is apparent that the reorganization signifies a new approach to current and future ER & D programs. Proposals by the AEC in response to President Nixon's directive of June 29, 1973, have been under close scrutiny.

AEC's recommended five-year national ER & D program is summarized in Table 1.

In November 1973, shortly after the onset of the Arab oil embargo, President Nixon proclaimed Project Independence and declared that "by the end of this decade, we will have developed the

Table 1
AEC'S RECOMMENDED NATIONAL ENERGY RESEARCH AND DEVELOPMENT PROGRAM, FISCAL YEARS 1975–79
(\$ millions)

| Self-Sufficiency Task | Estimated Costs to Attain— | | | |
	Short-term objectives (1975-85)	Mid-term objectives (1986-2000)	Long-term objectives (beyond 2000)	Total
Conserve energy and energy resources	1,160	280		1,440
Increase domestic production of oil and gas	430	30		460
Substitute coal for oil and gas on massive scale	1,690	485		2,175
Validate the nuclear option	1,100	2,990		4,090
Exploit renewable energy sources to the maximum extent feasible	135	150	1,550	1,835
Total	4,515	3,935	1,550	10,000

Source: U.S. Atomic Energy Commission, *The Nation's Energy Future*, WASH-1281 (December 1, 1973), p. 8.

potential to meet our energy needs without depending on any foreign sources." Concurrently, the AEC claimed that self-sufficiency may be attained by 1985. A report by the National Academy of Engineering states that:

> For the country to be self-sufficient even by 1985 would require doubling coal production, raising nuclear-power capacity fivefold, increasing hydroelectric and geothermal power production, stepping up oil and gas production 1¼ times, and producing big amounts of synthetic gas and liquid fuels from coal and oil from shale.[2]

Although the AEC expected 45 percent of the $10 billion energy budget to produce results before 1985 (Table 1), this fact is now considered doubtful and unrealistic by many energy experts.

In Table 2, the AEC's energy program for the period 1975–79 is broken down into individual program elements. The breeder program accounts for 28 percent and by far the largest part of the recommended budget. Is it justifiable to assign top priority to a program which cannot begin to supply any energy until 1987, given that the nation will be facing an energy crisis for the next fifteen years? Should a portion of the scarce resources allocated to mid- and long-term programs be channeled back to short-term programs? A thorough reassessment of the breeder program is mandatory before it is decided to continue that program at an intensified level.

The rationale for the traditionally strong support of the breeder program is succinctly stated in President Nixon's Energy Message to the U.S. Congress of July 4, 1971: "Our best hope today for meeting the Nation's growing demand for economical clean energy lies with the fast breeder reactor."[3] Will a breeder reactor generate electricity at a lower total cost, as promised? Almost all of the nuclear power plants in operation, under construction, or on order are light-water reactors (LWRs), which convert about 1 to 2 percent of the energy available in uranium into heat and electricity. In sharp contrast, a breeder reactor can unlock more than 60 percent of the energy in uranium. But is that sufficient to compensate for its initially higher plant capital cost and its research and development (R & D) costs? This is the main topic of our study. There are obviously other aspects

---

[2] Les Gapay, "U.S. Self-Sufficiency in Energy by 1980 Is Unlikely, Experts Say," *Wall Street Journal*, June 30, 1974, p. 15.

[3] President Richard M. Nixon's Energy Message, quoted in U.S. Atomic Energy Commission, Division of Reactor Development and Technology, *Updated (1970) Cost-Benefit Analysis of the U.S. Breeder Reactor Program*, WASH-1184 (January 1972), p. 1.

Table 2

AEC'S RECOMMENDED NATIONAL ENERGY RESEARCH AND
DEVELOPMENT PROGRAM, BY INDIVIDUAL PROGRAM
ELEMENT, FISCAL YEARS 1975–79

($ millions)

Self-Sufficiency Task	Recommended Program, FY 1975–79
Conserve energy and energy resources	
Reduced consumption	210
Increased efficiency	1,230
	1,440
Increase domestic production of oil and gas	
Production	310
Resource assessment	150
	460
Substitute coal for oil and gas on a massive scale	
Mining	325
Direct combustion	200
Synthetic fuels	1,270
Common technology	380
	2,175
Validate the nuclear option	
Safety, enrichment, HTGR, and other	1,245.7
Breeder	2,844.3
	4,090.0
Exploit renewable energy sources to the maximum extent feasible	
Fusion	1,450
Solar	200
Geothermal	185
	1,835
Total	10,000

Source: USAEC, *Nation's Energy Future*, p. 15.

of the fast breeder reactor (FBR) worthy of consideration. For example, it is often claimed that the FBR can reduce the environmental impact on air and water. But at the same time, a larger amount of the highly toxic plutonium must be handled, shipped, fabricated, processed, and safeguarded. It is also claimed that the breeder program will help the United States to maintain its leadership in nuclear power.[4] Is such an objective compelling enough to justify the devel-

---

[4] USAEC, *Updated (1970) Cost-Benefit Analysis*, WASH-1184, p. 7.

opment of the FBR at the proposed intensity? Or can the same goal be achieved by placing more emphasis on the fusion reactor program, which has recently shown several breakthroughs?

A fusion reactor is considered to be the prime electrical energy generator of the next century. Its fuel is low cost and practically unlimited. Environmentally, it has the potential for major reduction in the release of waste heat. When compared with the FBR, it has lower radioactive release, is less susceptible to major accidents, and presents less severe problems in the storage of radioactive waste. Stephen O. Dean of the AEC stated recently that, with adequate funding, the demonstration fusion-power plant could be introduced as early as 1995, instead of 2000 as previously planned.[5] On the other hand, the capital cost of the fusion plant may be very high. Also the technological and the engineering development of the FBR are at a significantly more advanced stage than that of the fusion reactor. These are important issues, which, however, will not be specifically investigated in detail in the present study. Instead, we will determine whether the fast breeder reactor can provide the United States with lower-cost electricity, as the AEC and many others have often claimed affirmatively and emphatically.

Commercial Nuclear Reactors and the Liquid Metal Fast Breeder Reactor

There are presently three types of commercial nuclear reactors in the United States: the boiling-water reactor (BWR), the pressurized-water reactor (PWR), and the high-temperature gas reactor (HTGR). The first two are cooled with ordinary (light) water and are called light-water reactors (LWRs). The water also serves as the moderator to slow down the neutrons needed for succeeding fissions. The basic form of the BWR is shown in Figure 1. It is fueled with enriched (to about 3 percent) UO_2 sealed in zirconium tubing. The stream is produced in the reactor, which operates at a pressure of about 1,000 pounds per square inch, and is then directed to the turbine. The PWR, as shown in Figure 2, is also fueled with enriched UO_2, but it has two loops instead of one. The coolant stays inside the containment vessel. The steam for the turbine is produced in the second loop which receives heat from the primary coolant loop via a heat exchanger. The pressure in the primary loop is about 2,000 pounds per

---

[5] Stephen O. Dean, "Fusion Power: Types, Status, Outlook," *Power Engineering*, March 1974, pp. 46-49.

Figure 1
BOILING-WATER REACTOR

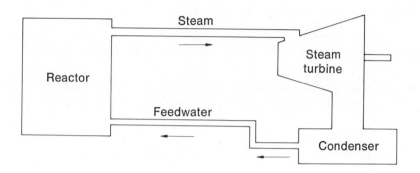

Source: Timothy J. Healy, *Energy, Electric Power and Man* (San Francisco, Cal.: Boyd and Fraser, 1974), p. 137.

Figure 2
PRESSURIZED-WATER REACTOR

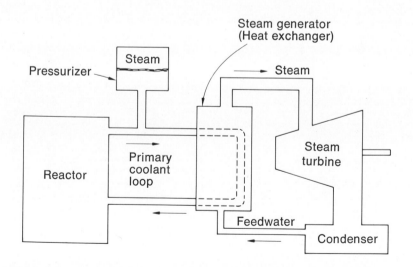

Source: Healy, *Energy, Electric Power and Man,* p. 137.

square inch. Since the coolant is confined in the primary loop, the chance for radioactive release is reduced. But the BWR operates at lower pressure and, in case of accident, the boiling water in the reactor becomes a poorer moderator and slows down the chain reaction as it converts into steam. Therefore, the BWR is inherently safer. Both types of reactors have a thermal efficiency of approximately 32 percent, and they can only utilize 1 to 2 percent of the energy potentially available in natural uranium. Natural uranium contains 0.711 percent of uranium–235, 99.283 percent of uranium–238, and 0.006 percent of uranium–234. Uranium–235 is fissile while uranium–238 is fertile. Only fissile material can be used in a nuclear reactor to sustain a chain fission reaction and release energy. But a fertile nucleus can convert itself into a fissile one by capturing a neutron. The aforementioned 1 to 2 percent fuel utilization includes the energy from fissile plutonium–239 which is produced in the reactor from uranium–238. It is thus apparent that an LWR is very inefficient in generating fissile materials from the abundant and fertile uranium–238.

The high-temperature gas reactor (HTGR) uses a gas coolant such as helium, which is heated to approximately 1400°F. Its basic form is shown in Figure 3. The steam is produced at a high temperature of 1000°F, compared with 600°F in the LWR, achieving an efficiency of 40 percent. In addition, the HTGR can extract 4 to 5 percent of the total available energy from thorium-uranium fuel instead of only 1 to 2 percent extracted by the LWR. An HTGR is initially fueled by uranium–235, with thorium–232 as the fertile material which converts into fissile uranium–233 in the reactor during operation. In subsequent reloads, the bred uranium–233 is used to reduce the amount of enriched uranium–235 in the fuel. Because of the HTGR's higher thermal efficiency and its utilization of thorium, the demand on uranium would be cut by half if all nuclear power plants were of this type. A 1000-MWe HTGR uses annually only eight tons of ThO_2, a form of thorium in the ore. The thorium reserve, as shown in Table 3, is sufficient to support even a predominantly HTGR power supply system for at least up to the year 2050, the terminal year used in the AEC's cost-benefit analyses. The capital costs for small units of HTGRs are higher than for those of LWRs. However, when the CONCEPT computer code was used to estimate the HTGR cost and to scale the costs to larger units, 1300 MWe and 2000 MWe, the costs were essentially the same as those for LWRs of the same size.[6]

---

[6] U.S. Atomic Energy Commission, *Proposed Final Environmental Statement, Liquid Metal Fast Breeder Reactor Program*, WASH-1535 (December 1974), vol. 4, p. 11.2-81.

Figure 3
HIGH-TEMPERATURE GAS-COOLED REACTOR

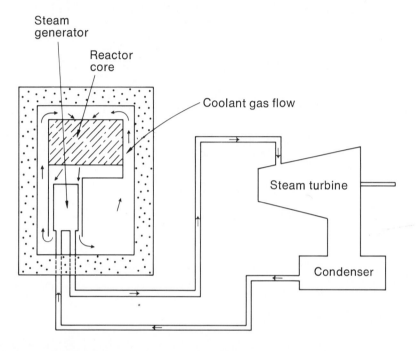

Source: Healy, *Energy, Electric Power and Man,* p. 138.

Table 3
ESTIMATED U.S. THORIUM RESOURCES
(thousand tons)

Cutoff Cost ($ per lb.)	Thorium Resources		
	Reasonably assured	Estimated additional	Total
10	65	335	400
30[a]	200	400	600
50[a]	3,200	7,400	10,600

[a] Includes lower-cost resources.

Source: U.S. Atomic Energy Commission, *Proposed Final Environmental Statement, Liquid Metal Fast Breeder Reactor Program,* WASH-1535 (December 1974), vol. 3, pp. 6A. 1-15 and 6A. 1-87.

Contrary to AEC opinion, HTGR capacity will certainly exceed LWR capacity in a nonbreeder power system.

Presently, there are five major manufacturers of reactors in the United States. The two leaders are Westinghouse and General Electric, which together have nearly 70 percent of the market. They are followed by Combustion Engineering, Babcock and Wilcox, and General Atomic (a joint venture between Gulf and Shell). General Electric builds only BWRs, and General Atomic only HTGRs, while the other three manufacturers produce PWRs.

The presently available commercial reactors are all "burners." Though they convert a small fraction of the fertile material into fissile isotopes, the bred material is insufficient to refuel the reactor. Thus, a breeder reactor must have a breeding ratio exceeding one in order for it to produce more fissile material than it consumes. There are several breeder reactors in different developmental stages, such as the light-water breeder reactor (LWBR), the molten-salt breeder reactor (MSBR), the gas-cooled fast reactor (GCFR), and the liquid metal fast breeder reactor (LMFBR). The LMFBR has been best funded and developed, and holds the greatest potential for early commercial use. Other nations with significant breeder reactor programs, such as the U.S.S.R., Great Britain, France, Germany, Italy, and Japan, have also made the same choice. A schematic diagram of the LMFBR is shown in Figure 4.

Fast neutrons are needed to breed fissile materials efficiently, and since water is a neutron moderator, it cannot be used as a coolant. Instead, liquid sodium is used. Plutonium–239 is the fuel, and the excess neutrons emitted by plutonium are used to convert the fertile uranium–238 to fissile plutonium–239. Since uranium–238 is 140 times more abundant in nature than uranium–235, the potential life of nuclear fission reactors as a power source can be increased by about a hundredfold. But, do we need fission reactors to last for thousands of years when alternative energy resources such as nuclear fusion and solar energy will certainly be competitive, if not conclusively superior, within only decades?

The Liquid Metal Fast Breeder Reactor Program in the United States

A short history of the LMFBR program will indicate the level of our involvement.[7] The program can be traced back to at least 1945, when a project to develop the plutonium-fueled FBR was initiated in the

---

7 Ibid., vol. 1, pp. 2.2-1 to 7.

Figure 4

LIQUID METAL FAST BREEDER REACTOR

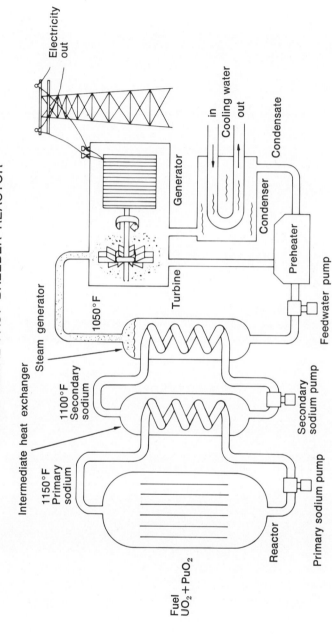

Source: Gerard M. Crawley, *Energy* (New York: Macmillan, 1975), p. 130.
© 1975 by Gerard M. Crawley.

11

Argonne National Laboratory Division of the Manhattan District Metallurgical Laboratory. The construction of the Clementine reactor at Los Alamos began in 1946. Until its shutdown in 1953, it was used to explore the possibility of constructing a fast reactor utilizing plutonium fuel, fast neutrons, and a liquid metal coolant, which was mercury at that time. Design work on the Experimental Breeder Reactor I (EBR-I) began in 1945 and actual operation began in 1951. Until its shutdown in 1963, it had demonstrated the feasibility of breeding and of using liquid metals (such as sodium-potassium) as coolants and the possibility of designing an inherently stable fast breeder. In 1954, the AEC gave the development of nuclear power, including the fast breeder program, a great boost by initiating the Five-Year Reactor Development Program. However, in the following decade, major efforts of national and industrial laboratories were focused on LWRs where the prospect of large-scale commercial application was nearer at hand. The Experimental Breeder Reactor II (EBR-II), whose design, in addition to the reactor, includes a reprocessing and fabrication system for fuel recycling, began in 1956 and went critical in 1963. It has been providing vital information on the effect of fast neutron irradiation on various materials and is expected to continue in operation. The Enrico Fermi Fast Breeder Reactor achieved criticality in 1963 and was the first privately owned breeder. However, after generating 32 million kilowatt-hours of electricity, the project was terminated in 1972 due to lack of funds.

In addition to the above reactors, the Los Alamos Molten-Plutonium Reactor Experiment (LAMPRE) was in operation during 1961–64 to investigate the use of molten fuel, and the Southwest Experimental Fast Oxide Reactor was in operation during 1969–72 to demonstrate the inherent safety of LMFBRs using mixed-oxide fuel.

Another significant boost to the breeder program occurred in 1967 when the AEC issued a supplement to the 1962 report to the President on civilian nuclear power and reaffirmed its promise to assign the LMFBR program the highest priority. The principal reason for promoting LMFBR as the top-priority program was the AEC's scaled-down involvement in LWRs after the utility industry began to make heavy commitments for LWRs in 1965. As breeder reactors can provide the most efficient fuel utilization, the stage was set for a scaled-up breeder reactor program. As is evident from Table 4, the budget for the LMFBR program increased significantly in 1967.

The design of the Fast Flux Testing Facility (FFTF) began in 1967 and is currently under construction at the Hanford Engineering Development Laboratory. It is scheduled for completion in 1977 and

Table 4

EXPENDITURES ON THE LMFBR PROGRAM, 1947–79
($ millions)

Fiscal Year	Annual Expenditure	Cumulative Expenditures
1947	0.9	0.9
1948	0.6	1.5
1949	1.1	2.6
1950	2.4	5.0
1951	0.7	5.7
1952	3.5	9.2
1953	3.4	12.6
1954	3.9	16.5
1955	5.0	21.5
1956	5.8	27.3
1957	7.6	34.9
1958	16.1	51.0
1959	21.8	72.8
1960	20.3	93.1
1961	14.3	107.4
1962	17.1	124.5
1963	26.0	150.5
1964	26.7	177.2
1965	34.1	211.3
1966	42.6	253.9
1967	78.0	331.9
1968	95.0	426.9
1969	106.0	532.9
1970	120.0	652.9
1971	176.0	828.9
1972	205.0	1,033.9
1973	253.8	1,287.7
1974	356.8 [a]	1,644.5
1975	477.0 [b]	2,121.5
1976	538.6 [b]	2,660.1
1977	510.8 [b]	3,170.9
1978	524.2 [b]	3,695.1
1979	506.0 [b]	4,201.1

[a] Planned.

[b] Recommended.

Source: Data prior to 1973: U.S. Atomic Energy Commission, Division of Reactor Development and Technology, *Liquid Metal Fast Breeder Reactor Program Plan,* WASH-1101, 2nd ed. (December 1973), p. 8; data for 1973-79: USAEC, *Nation's Energy Future,* p. 29.

will be used to test fuel and other reactor core components for both demonstration and future commercial plants.

Another major facility is the LMFBR Demonstration Plant. Three proposals for its construction were received by the AEC in late 1971. After negotiations in 1972, the AEC reached a contractual arrangement with the Tennessee Valley Authority and Commonwealth Edison on July 25, 1973, to develop, design, construct, and operate the plant. In March 1972, two nonprofit and tax-exempt corporations were formed—the Breeder Reactor Corporation (BRC) and the Project Management Corporation (PMC). The BRC coordinates the financial aspects relating to the electric utilities' participation in the plant and acts as liaison between the project and the utilities. The PMC is the central management and contracting organization for the project. The AEC directed and administered all technical activities of the Nuclear Steam Supply System and, of course, provided support from the LMFBR development program. In August 1972, a plant site was selected on the Clinch River in East Tennessee. But so far not a spade of earth has been lifted and the project is now at least two years behind schedule.

This demonstration plant will test the functioning of plant components and provide information on the operation of an integrated large-scale power system, including its economic and environmental impact and its degree of safety and reliability. The electrical power will be 350 MWe. The estimated construction cost of the plant has increased from $700 million to $1.7 billion since the project was announced only three years ago,[8] and, so far, the utility industry has pledged about $260 million as its share. The construction and operation of this plant is essential for the eventual use of commercial LMFBRs. The question is whether only one demonstration plant at 350 MWe will be sufficient to justify commercial introduction of 1300-MWe reactors in 1987. If not, or if the initial commercial LMFBRs have to be federally subsidized, additional costs will be incurred. Recently the AEC estimated that a subsidy of only $276 million would be needed for the construction of a Near Commercial Breeder Reactor (NCBR).[9]

At this juncture it will be useful to look to the past and to the future in an effort to reassess the LMFBR program and its effect on the nation.

---

[8] Ibid., vol. 4, p. 11.2-35.

[9] Ibid., p. 11.2-33.

Cost-Benefit Analyses by the Atomic Energy Commission

The initial results of the first extensive cost-benefit analysis of the U.S. Breeder Reactor Program by the AEC were available in 1968 and the report was published in 1969 (hereafter referred to as CB-68).[10] The initial findings were updated in 1970, finalized in 1972 (CB-70),[11] and again updated and released in March 1974 (CB-74D)[12] as part of the *Draft Environmental Statement, Liquid Metal Fast Breeder Reactor Program.* Recently, the analysis in the draft was slightly modified, expanded to include additional cases (CB-74),[13] and appeared in the *Proposed Final Environmental Statement, LMFBR Program.*

The figures given in the conclusions of these analyses vary widely, reflecting the uncertainties of the projections used in the analyses and the sensitivity of the results to these projections. However, the principal conclusion claimed by the AEC remains unaltered, namely, that "the introduction of a fast breeder into the U.S. electric power utility system will produce significant financial benefits. This results largely from a reduction in uranium ore and U-235 enrichment requirements."[14] Table 5 shows the net discounted benefits for the base cases in the analyses. Variations are due principally to changes in uranium and fossil fuel prices, energy demand, plant capital costs, rate and date of LMFBR introduction, constraints on plant mix, separative work charge, environmental costs, R & D costs, and computer programs. Were these projections reasonably chosen? How would the benefit from FBR be affected if several of the projections were adjusted simultaneously? In response to the comments of the Environmental Protection Agency (EPA), the Natural Resources Defense Council (NRDC), and others, the AEC has included many more cases in its latest analysis. It makes our task of evaluating the AEC analysis much less formidable.

This study seeks to evaluate the projections and assumptions employed in the AEC analyses, particularly those in CB-74, and to indicate how the results would be modified if the parameters were adjusted according to our proposals. Some of the modifications have already been calculated by the AEC in its sensitivity analysis. The

---

[10] U.S. Atomic Energy Commission, Division of Reactor Development and Technology, *Cost-Benefit Analysis of the U.S. Breeder Reactor Program*, WASH-1126 (April 1969).

[11] USAEC, *Updated (1970) Cost-Benefit Analysis*, WASH-1184.

[12] U.S. Atomic Energy Commission, *Draft Environmental Statement, Liquid Metal Fast Breeder Reactor Program*, WASH-1535 (March 1974).

[13] USAEC, *Proposed Final Environmental Statement*, WASH-1535.

[14] Ibid., p. 11.3-1.

Table 5

LMFBR BENEFITS ACCORDING TO THE AEC'S ANALYSES
(billions of current dollars)

	Net Discounted Benefits	
Analysis	Discount rate of 7 percent	Discount rate of 10 percent
CB-68	6.6	0.0
CB-70	19.1	4.3
CB-74D	55.5[a]	18.2
CB-74	48.6[a]	14.7

[a] At 7.5 percent discount rate.

Source: USAEC, *Cost-Benefit Analysis*, WASH-1126, pp. 43-44; USAEC, *Updated (1970) Cost-Benefit Analysis*, WASH-1184, pp. 29-30; USAEC, *Draft Environmental Statement*, WASH-1535, vol. 3, pp. 3-8 and 3-11; and USAEC, *Proposed Final Environmental Statement*, WASH-1535, vol. 4, pp. IV.D-1 and IV.D-16.

modifications are not necessarily caused by events which occurred since the completion of the analyses. Some of the assumptions used were inappropriate even under the economic environment existing at the time of the early analyses.

All of the AEC analyses start with a projection of electrical energy demand to the year 2020. The cost of electricity without and with the integration of the LMFBR into the electric power system is then calculated by employing a linear program with an objective function minimizing the total electric cost over the planning horizon. The difference between the two energy costs, with LMFBR program costs subtracted, is the net gross benefit. The net discounted benefit is the present worth of the net gross benefit. Treating electrical energy demand independently of electricity cost is fundamentally unsound because it assumes the demand of electricity to be absolutely inelastic.

Let us assume the demand curve for electricity to be DD' as shown in Figure 5. If the cost per unit of electricity without LMFBR is OP_0 and that with LMFBR is OP_1, the saving of electricity cost before LMFBR program costs should be equal to the shaded area P_1P_0ABC, which is the difference between the consumers' surpluses, P_1DB and P_0DA The AEC only considers the rectangular area P_1P_0AC in its analyses. The consumer surplus created by the additional electricity used, represented by triangle ABC, is ignored. In the case shown in Figure 5, the AEC underestimates the benefit from the

Figure 5

DEMAND CURVE OF ELECTRICITY

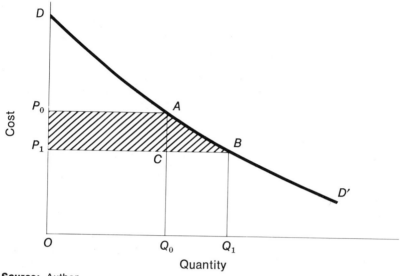

LMFBR. On the other hand, if OP_1 is greater than OP_0, the AEC can underestimate the loss from the LMFBR also. In its future analyses, the AEC should attempt to project and consider the effect of the elasticity of electricity demand.

Another aspect to be considered is synergistic uncertainty. Since each assumption is associated with a degree of uncertainty, the result derived from a combination of assumptions can be highly uncertain. The AEC should present a risk analysis similar to those frequently used for capital investment.[15] First it is necessary to estimate the likelihood of occurrence of each value of the parameters in each assumption. The final result can be represented by a probability curve of the rate of return, as shown in Figure 6. It shows the probability of any rate of return and the variances, and can be compared with curves for alternative programs.

---

[15] David B. Hertz, "Risk Analysis in Capital Investment," *Harvard Business Review*, vol. 42, no. 1 (January/February 1964), pp. 95-106.

Figure 6

PROBABILITY CURVE OF THE RATE OF RETURN

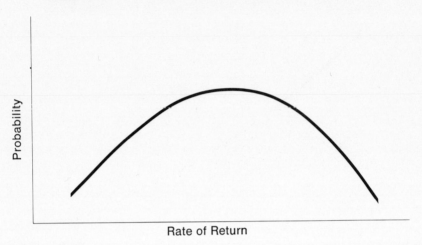

Source: Author.

Summary of the Findings of this Study

The findings of this study can be summarized in the flow chart shown in Figure 7. The arrow emerging from the top block signifies that the net discounted benefit of the LMFBR for the base case (Case 3) in CB-74 is $14.7 billion. The discussion on discount rate can be found in Chapter 2. The benefit is reduced to $7.3 billion or $6.8 billion, depending on the projection of uranium supply-versus-price schedule. Cases that have been calculated by the AEC are shown with their case numbers below the dollar figures.

If we adopt any reasonable set of assumptions, none of the cases will show a net discounted benefit from the LMFBR. The AEC's repeated assertion that the LMFBR can provide an overall saving on the cost of electric energy is not substantiated here. The AEC's major assumptions are investigated in Chapters 2 through 5. The study's conclusions are set forth in Chapter 6.

For convenience of comparison, all dollar figures, unless specified otherwise, are stated in mid-1974 dollars and, if discounted, are discounted to mid-1974.

18

Figure 7
SUMMARY OF THE FINDINGS OF THIS STUDY

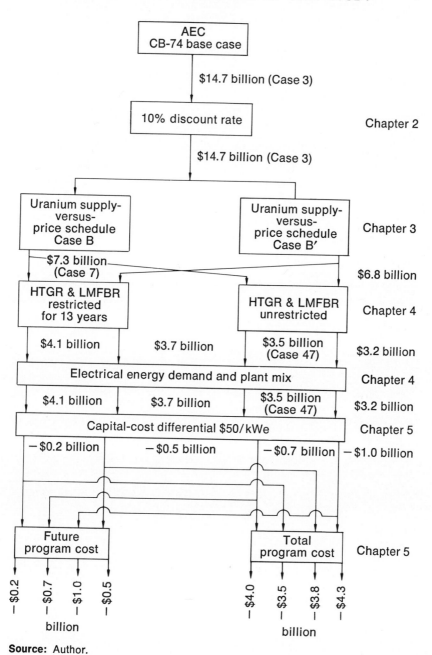

Source: Author.

2
DISCOUNT RATE

Discount Rates Used by the Atomic Energy Commission

The costs and benefits of a project are seldom incurred in an identical stream with respect to time and quantity. They must be discounted at a rate to the present in order to reflect time-preference and time-productivity. Time-preference refers to one's willingness to exchange units of current cost for units of future benefit, while time-productivity denotes one's ability to convert units of current investment into units of future return. An appropriate discount rate is essential if resources are to be utilized in the most efficient way. A too high rate would discourage worthwhile long-term projects and a too low rate would divert resources from the urgent needs of the nation. In two of the AEC analyses, CB-68 and CB-70, a rate of 7 percent was considered to be the most appropriate. The rate was raised to 7.5 percent in CB-74D. In CB-74, the AEC reluctantly chose a 10 percent rate, while still considering a 7.5 percent or lower rate to be more appropriate. A change of discount rate from 7.5 percent to 10 percent reduces the net discounted benefit of the LMFBR from $48.6 billion to $14.7 billion. This drastic reduction resulted from the continuous cost outflow in the near term and no benefit inflow until the 1990s.

The discount rates used in all the AEC cost-benefit analyses were derived from the after-tax return of utilities. The philosophy was clearly stated in CB-68:

> The LMFBR program can be identified with the utility sector of the U.S. economy, and the rate of return applicable to that sector has been considered as the criterion rate for evaluation of public investments in this area. The discount rates appli-

cable to the electric utility industry would most nearly comply with this criterion.[1]

Such an argument is at fault because the government should carry out programs to best benefit society, not any particular industry. The rate of return of utilities is a regulated rate, and other industries have different rates of return. Let us assume that industry A has a zero rate of return and industry B a 15 percent rate of return. If government investments are based on these rates, resources will be channeled to industries with the lower rate of return, thereby resulting in government subsidization of the less efficient industries. Society will suffer because the overall rate of return from all government projects will be reduced.

Using an after-tax rate of return is also inappropriate because the tax paid by the utilities is a portion of the benefit returned to society. The AEC should have used a before-tax rate, and computed both costs and benefits before taxes. The choice of discount rate should be based on the two theories summarized in the next section.

Discount Rate Determination

One of the two theories has been succinctly stated by William J. Baumol:

> The correct discount rate for the evaluation of a government project is the percentage rate of return that the resources utilized would otherwise provide in the private sector. That is, the correct discount rate is the opportunity cost in terms of the potential rate of return in alternative uses on the resources that would be utilized by the project.[2]

Therefore, resources should not be transferred to a government project if those resources, left in private hands, can produce a higher rate of return to benefit society. If the resources originate from different sectors, the rate of return is the weighted average of the rates of those sectors. In practice, it is often a difficult task to quantify the amount of resources extracted from various sectors. The LMFBR program will continue to draw resources from other types of nuclear reactors, as well as from equipment manufacturers and electric utilities. The precise proportions from each sector are hard to determine.

---

1 USAEC, *Cost-Benefit Analysis*, WASH-1126, p. 38.
2 William J. Baumol, "On the Discount Rate for Public Projects," *The Analysis and Evaluation of Public Expenditure*, 91st Congress, 1st session, 1969, vol. 1, p. 491.

Jacob A. Stockfisch estimated that the annual average rates of return in the period 1961 to 1965 were 15.4 percent for the manufacturing industry and 9.3 percent for the electric utilities.[3] Stockfisch also estimated the average rate of return before taxes in the private sector to be 13.5 percent while George J. Stigler of the National Bureau of Economic Research estimated a rate of 14 percent.[4] The variation in rates resulted from choice of time period and weighting. Paul W. MacAvoy stated in his in-depth study on the FBR program that "the actual rate cannot be less than 10 percent per annum and might well not be greater than 15 percent."[5] Thus, according to the above theory, the discount rate used for the breeder reactor program has to be at least 10 percent.

The second theory for the determination of discount rates is based on the concept of opportunity cost of public investment or borrowing. At least two approaches can be taken to quantify the concept. In hearings before the Subcommittee on Economy in Government of the Joint Economic Committee, Otto Eckstein suggested using the taxation approach, which yielded a discount rate of approximately 8 percent in the economic environment of 1968.[6] The long-term government bond average rose from 5¼ percent in 1968 to around 6¾ percent in 1974. The interest rate structure therefore moved up about 1½ percentage points in that period. Eckstein's approach applied to the economy of 1974 would yield a discount rate of approximately 9½ percent. Since the historic trend of interest rates is upward, interest rates are not expected to return to the 1968 level on a permanent basis. The second approach to quantify the theory is based on the cost of government borrowing, an approach often followed by federal agencies for discount rate guidelines. The following guideline was proposed in 1968:

> The interest rate to be used in plan formulation and evaluation for discounting future benefits and computing cost, or otherwise converting benefits and costs to a common time basis, shall be based upon the average yield during the preceding fiscal year on interest-bearing marketable securities

---

[3] Statement of Jacob A. Stockfisch, quoted in U.S. Congress, Joint Economic Committee, *Interest Rate Guidelines for Federal Decision Making*, 90th Congress, 2d session, January 1968, p. 7.

[4] Statement of George J. Stigler, quoted in ibid., p. 62.

[5] Paul W. MacAvoy, *Economic Strategy for Developing Nuclear Breeder Reactors* (Cambridge: M.I.T. Press, 1969), p. 92.

[6] Statement of Otto Eckstein, Hearings before the Subcommittee on Economy in Government, Joint Economic Committee, *Economic Analysis of Public Investment Decisions: Interest Rate Policy and Discounting Analysis*, 90th Congress, 2d session, 1968, pp. 50-57.

of the United States which at the time the computation is made, have terms of 15 years or more remaining to maturity. Provided, however, that in no event shall the rate be raised or lowered more than one-quarter percent for any years.[7]

The comptroller general of the United States carried the proposal one step further. He advocated the inclusion of an adjustment factor to account for the taxes foregone by the government. Arnold C. Harberger endorsed this suggestion, but argued that the adjustment factor should also include the property, excise, and sales taxes foregone. He stated:

> From the point of view of economic logic, when attempting to measure the cost to society as a whole of Government borrowing, there is no ground to distinguish whether the taxes foregone on the income from displaced investments would have accrued to the Federal Government or to State and local governments. By focusing on Federal taxes alone, the report unduly narrows its focus. Property tax considerations alone must enter into any comprehensive measure of the overall opportunity cost of Government borrowing, particularly when property taxes play as important a role in the overall fiscal system as they do in the United States.[8]

He went on to comment on the inclusion of excise and sales taxes:

> . . . even apart from highway taxes which might be considered as user charges, the excises by themselves produce some $10 billion per year in revenue to the Federal Government. Of course, State and local sales and excise taxes should be brought into account; these, which yield considerably more than the Federal excises, further emphasize the importance of incorporating sales and excise taxes into the overall calculation of the opportunity cost of Government borrowing.[9]

We are now ready to compute the discount rate in accordance with the above guidelines. First, we calculate the discount rate that would have been used in CB-74 had the comptroller general's guidelines been followed.[10]

---

[7] Statement of Henry P. Caulifield, Jr., in ibid., p. 13.

[8] Statement of Arnold C. Harberger, in ibid., p. 60.

[9] Ibid.

[10] Joint Economic Committee, *Interest Rate Guidelines*, pp. 60-63.

Average interest rate on long-term government bonds [11] during the first half of 1974	6.75 percent
Corporate taxes foregone assuming:	3.51 percent

(1) Average corporate return on investment before taxes—13.5 percent,

(2) Marginal corporate tax rate—40 percent,

(3) Fraction of dollars borrowed by the government which would have gone into corporate investment—65 percent

Personal taxes foregone assuming:	1.05 percent

(1) Average return on proprietorship, personal income—producing investments and etc. before taxes—10 percent

(2) Composite marginal tax rate—30 percent

Adjustment:	1.12 percent

Due to taxes foregone on dividends assuming taxable dividend payout of corporate profits after taxes to be 40 percent, and

Due to personal taxes foregone by the government if the corporate investment is financed by bonds instead of earnings, assuming the interest rate for corporate bonds to be 8.25 percent

Subtraction of income taxes collected on government interest payments assuming the composite tax rate for corporations and individuals to be 35 percent	(2.36 percent)
Cost to government	10.07 percent

Therefore, it is justifiable to use a discount rate of 10 percent in CB-74. Following the same procedure, a discount rate of 9.3 percent instead of 7 percent should have been used in CB-70. The above consideration has not even included an adjustment for property, excise, and sales taxes foregone—as proposed by Harberger, who estimated that the appropriate discount rate in 1968 was approxi-

---

[11] Long-term government securities are those that are due or callable after ten years or more. If the average interest rate is based on securities having terms of fifteen years or more remaining to maturity as specified in the guideline, the interest rate would be even somewhat higher.

mately 10.7 percent.[12] If Harberger's estimate were made today, the discount rate would be approximately 11.4 percent. Inflation is the strongest argument in support of a significantly lower discount rate. Interest rates should be adjusted downward for inflationary expectations if one expresses the cost and benefit streams in constant dollars. Henry P. Caulifield, Jr., expressed this point of view forcibly before the Subcommittee on Economy in Government.[13] Harberger's reply in the same hearings was this:

> The extent of inflationary anticipations in our present structure is very minor in comparison with the differences between the Government bond rate and the yield of capital in the private economy, and, to put it another way, very minor in comparison with the weight of the tax adjustments that the Comptroller General's report and my own statement consider to be the major component of the difference.[14]

Stockfisch did estimate the real average rate of return in the private sector for the period from the Korean War to the Vietnam War by subtracting the average rate of inflation from the aggregate average rate and arrived at a rate of 10.4 percent.[15]

A discount rate derived from the rate of return on investment in the private sector is higher than one computed from the cost of government borrowing. The former is a discount rate subject to risk, while the latter is usually considered the riskless rate. The fact that the government can carry on a large number of projects simultaneously allows that these projects, on a whole, be considered "riskless" in a statistical sense. However, the risk of a particular project under consideration should be accounted for. For example, let us assume that there are two government projects—one with high risk and the other with low risk—and that both of them give identical cost and benefit streams. Without the inclusion of a risk factor to account for the risk above the average risk of all government projects either explicitly in the discount rate or implicitly in the calculation of costs and benefits, what is the criterion for choosing the low-risk project?

In March 1972, George P. Shultz, then director of the Office of Management and Budget, directed all agencies of the executive branch of the federal government (except the U.S. Postal Service) to use a

---

[12] Statement of Harberger, in Joint Economic Committee, *Interest Rate Policy*, p. 65.

[13] Statement of Caulifield, in ibid., pp. 46-47.

[14] Statement of Harberger, in ibid., p. 70.

[15] Jacob A. Stockfisch, *Measuring the Opportunity Cost of Government Investment* (Arlington, Va.: Institute for Defense Analyses, March 1969), p. 490.

10 percent discount rate before tax and after inflation in analyzing program proposals submitted to his office.[16] In fact, the Department of Defense has been using a 10 percent rate for all its military construction programs for a number of years.[17] In December 1971, the Water Resources Council suggested "that the appropriate rate for evaluating Government investment decisions is approximately 10 percent and is substantially invariant to short-term changes in economic and money market conditions."[18] The Soviet Union has also been using a 10 percent discount rate by and large.[19]

In conclusion, the 7 percent or 7.5 percent discount rate used in the AEC's cost-benefit analyses except for CB-74 was definitely too low even in the economic environment at the time when these analyses were conducted. The 10 percent rate used in CB-74 is appropriate but certainly not a conservative choice, as claimed by the AEC. Somewhat higher rates are also justifiable. The AEC should use the 10 percent rate as the before-tax, rather than the after-tax, rate, and compute the future benefit streams before tax so that the benefits of AEC projects can be compared with other governmental projects on an equitable basis. Using 10 percent as the after-tax rate is by no means more conservative. This is true if the project is expected to produce a net discounted benefit. On the other hand, if the project will produce a net discounted loss, which may very well be the case for the LMFBR, using an after-tax rate instead of a before-tax rate will understate the loss of the project.

---

[16] Statement of George P. Shultz, director, Office of Management and Budget, to the heads of executive departments and establishments, Circular No. A-94, revised, March 1972.

[17] Joint Economic Committee, *Interest Rate Guidelines*, p. 57.

[18] Water Resources Council, "Proposed Principles and Standards for Planning Water and Related Land Resources," 36FR245 (Washington, D. C., December 1971).

[19] Statement of Eckstein, in *Interest Rate Policy*, p. 73.

3
URANIUM SUPPLY SCHEDULES

Uranium Resources in the United States

The most distinct feature of the FBR is its potential for much better utilization of uranium fuel compared to the present type of commercial reactors, such as the pressurized-water reactor (PWR), boiling-water reactor (BWR), and the high-temperature gas reactor (HTGR). The cost of electricity in present commercial reactors is, however, only weakly dependent on the price of uranium ore, U_3O_8. Thus, the pertinent factor is the availability of uranium ore at various prices in the future. Uranium resources are often estimated in two categories—reserves and potential resources—which, according to the AEC, are defined as follows:

> Reserves—an estimate of the quantity of uranium in known deposits which can be produced at or below a stated cost per pound termed the "cutoff cost." The quantity, grade and physical characteristics have been established with reasonable certainty by detailed sampling, usually by surface drilling, initially and later supplemented by underground drilling and sampling. The term "reserves" is roughly synonymous with "reasonably assured resources" used by the International Atomic Energy Agency, and the two terms are often used interchangeably.
>
> Potential Resources—those surmised to occur in unexplored extension of known deposits, or in undiscovered deposits within or adjacent to uranium areas, or in other favorable areas. They are expected to be discoverable or exploitable at a given cost. The term is essentially synony-

Table 6

U.S. REASONABLY ASSURED AND POTENTIAL URANIUM RESOURCES

(cumulative thousands of tons of U_3O_8)

U_3O_8 Cost Per Pound [a]	1962	1966	1967	1968	1969	1970	1971	1973
(1) Reasonably Assured Resources								
$ 8.00	—	145	141	148	161	204	243	280
10.00	380	190	200	310	320	340	390	340
15.00	—	340	350	460	470	500	580	520
30.00	780	510	520	660	670	750	750	700
50.00	5,800				6,000		4,750	
100.00	12,000				12,000		8,750	
500.00	500,000							
(2) Potential Resources								
$ 8.00	—	—	—	—	225	390	490	450
10.00	420	325	325	350	350	600	680	700
15.00	—	525	525	550	550	950	1,040	1,000
30.00	720	965	965	990	1,000		1,660	1,700
50.00	3,700				4,000		3,660	
100.00	13,000				13,000		8,660	
500.00	1,500,000							

	(3) Total of Reasonably Assured and Potential						
$ 8.00	—	—	—	—	386	594	733	730
10.00	800	515	525	660	670	940	1,070	1,040
15.00	—	865	875	1,010	1,020	1,450	1,620	1,520
30.00	1,500	1,475	1,485	1,650	1,670		2,410	2,400
50.00	9,500				10,000		8,410	
100.00	25,000				25,000		17,410	
500.00	2,000,000							

a Production cost, not buyer's price; in current dollars.

Sources: USAEC, *The Nuclear Industry*, WASH-1174 (1966-71 and 1973); U.S. Congress, Joint Committee on Atomic Energy, *Nuclear Power Economics—1962 through 1967*, February 1968, p. 200; USAEC, *LMFBR Demonstration Plant Program*, WASH-1201 (March 1972), p. 197; USAEC, *Potential Nuclear Power Growth Patterns*, WASH-1098 (May 1973); and USAEC, Division of Production and Materials Management, *Nuclear Fuel Supply*, WASH-1242 (May 1973).

mous with "estimated additional resources" used by the International Atomic Energy Agency.[1]

Domestic uranium resources as estimated by the AEC's Division of Raw Materials are given in Table 6. The estimates of reasonably assured resources at less than $10 per pound are accurate to within around 20 percent. Estimates at higher cutoff costs are progressively less accurate. It should be apparent from the table that estimated resources have been revised upward even before the figures are expressed in constant dollars. Figure 8 compares the supply schedules used in CB-74 with those used in CB-68 and CB-70 and also shows the estimates of U.S. reasonably assured and potential uranium resources and of uranium requirements. The AEC considers the schedule for the base case to be the most reasonable estimate of uranium prices. It is apparent that the uranium prices used in the base case of CB-74 are significantly higher than those of previous analyses. It is inconsistent with the fact that U.S. resource estimates in 1973 or 1974, as shown in Table 6, are not any lower than the estimates made in 1971. Prices for the long term should be based on the estimates of uranium resources rather than on temporary price fluctuations. An overestimate of uranium prices leads to an overestimate of the LMFBR benefit. Even in the case which is considered to be optimistic by the AEC, in CB-74, uranium prices are still overestimated.

Correlation between Exploration Activities and Uranium Price

Past prospecting activities have been concentrated on ore priced at less than $10 per pound. There has been no extensive and systematic effort to explore uranium deposits at higher prices. Furthermore, exploration activities and additions to reserves are correlated with current as well as anticipated uranium prices, as is shown in Figure 9. In the mid-1950s, the large number of feet drilled and the large addition to reserves were correlated with the high price. The uranium price per pound declined from $17 in the mid-1950s to $6 in 1970 and has started to rise since then. A higher price and price expectation will certainly provide additional incentives for uranium development. It is not justifiable to assume drastically higher prices without assuming a corresponding increase in uranium reserves, as was the case in CB-74. There is little doubt that uranium resources will eventually exceed the present indication.

---

[1] U.S. Atomic Energy Commission, Grand Junction, Colorado, Office, *Nuclear Fuel Resource Evaluation: Concepts, Uses and Limitations,* GJO-105, May 1973.

Figure 8

URANIUM SUPPLY-VERSUS-PRICE SCHEDULES, URANIUM RESOURCES AND REQUIREMENTS

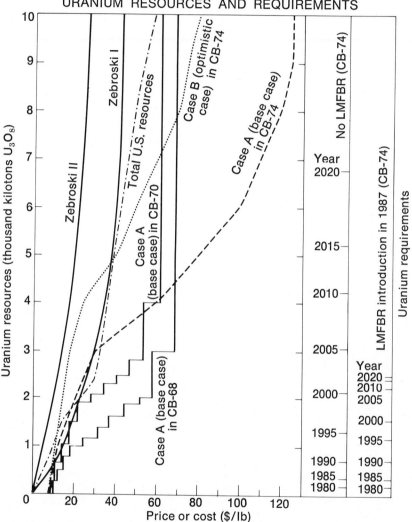

Source: United States Atomic Energy Commission, *Potential Nuclear Power Growth Patterns,* WASH-1098 (May 1973); United States Atomic Energy Commission, *Nuclear Fuel Supply,* WASH-1242 (May 1973); E. L. Zebroski, "Breeding— How Soon a Necessity?" *Nucleonics,* vol. 18, no. 2 (February 1960), p. 65; United States Atomic Energy Commission, *Updated (1970) Cost-Benefit Analysis of the U.S. Breeder Reactor Program,* WASH-1184 (January 1972), p. 42; United States Atomic Energy Commission, *Cost-Benefit Analysis of the U.S. Breeder Reactor Program,* WASH-1126 (April 1969), p. 70; and United States Atomic Energy Commission, *Proposed Final Environmental Statement, Liquid Metal Fast Breeder Reactor Program,* WASH-1535 (December 1974), vol. 4, pp. 11.2-72 and 11.2-144.

Figure 9

URANIUM PRICES AND ADDITIONS
TO URANIUM PROVEN RESERVES ($8/lb)

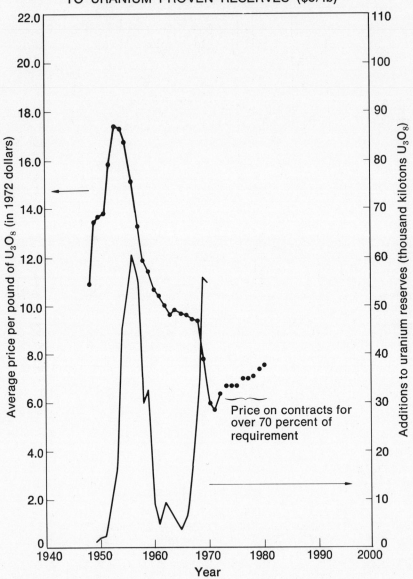

Price on contracts for over 70 percent of requirement

Source: Electric Power Research Institute, *Uranium Resources to Meet Long-Term Uranium Requirements,* SR-5, November 1974, p. 72; and Thomas B. Cochran, *The Liquid Metal Fast Breeder Reactor, An Environmental and Economic Critique* (Washington, D. C.: Resources for the Future, Inc., 1974), p. 91.

Underestimation of Present Uranium Resources

It should also be observed that uranium resources and costs are estimated mainly on the basis of existing proven exploration, mining, and
processing technology. New techniques in exploration and extraction
will be developed to provide larger quantities of uranium at lower
costs. To assume a practically exponential growth in future demand
for electricity and uranium but an essentially unimproved technology
is a dangerous proposition. Experience with the development of other
energy resources and metals teaches the same lesson. For example,
the estimated ultimate reserve of oil has increased five- to tenfold in
the last forty years.[2] Professor Edward Mitchell has provided an
excellent statement of this point:

> America has had less than a dozen years' supply of oil left
> for a hundred years. In 1866 the United States Revenue
> Commission was concerned about having synthetics avail-
> able when crude oil production ended; in 1891 the U.S. Geo-
> logical Survey assured us there was little chance of oil in
> Texas; and in 1914 the Bureau of Mines estimated total
> future U.S. production at 6 billion barrels—we have pro-
> duced that much oil every twenty months for years. Perhaps
> the most curious thing about these forecasts is a tendency
> for remaining resources to grow as we deplete existing re-
> sources. Thus, a geologist for the world's largest oil com-
> pany estimated potential U.S. reserves at 110 to 165 billion
> barrels in 1948. In 1959, after we had consumed almost
> 30 billion of those barrels, he estimated 391 billion were
> left.[3]

It should be mentioned that uranium deposits in the above-$50-
per-pound category include extensive granite bodies containing less
than fifty ppm of uranium. However, the possibility of cost reduction
due to new extraction techniques should not be overlooked.

Considering the trend of additions to uranium resources and assuming a long-term pattern of development similar to that of oil, E. L.
Zebroski suggested that the uranium supply-versus-price schedule
should lie between limits marked Zebroski I and II in Figure 8. The
lower curve is based on a twofold increase over the resources proven
and inferred in 1959. The upper curve reflects the projection of an
ultimately sevenfold increase over the estimates of 1959. Though his
observation was made in 1960, his approach is still valid today. In

---

[2] E. L. Zebroski, "Breeding—How Soon a Necessity?" *Nucleonics*, vol. 18, no. 2
(February 1960), p. 64.

[3] Mitchell, *U.S. Energy Policy*, p. 5.

1971 the uranium industry, in a report for the Nuclear Task Group of the National Petroleum Council, estimated uranium resources available at less than $15 per pound by extrapolating data on the abundance of uranium from explored to favorable unexplored areas. The report arrived at a figure several times higher than the AEC estimates.

Uranium Supply-versus-Price Schedules for this Study

Case B (the AEC's "optimistic" case) in CB-74 should be regarded as quite pessimistic, especially near the high-price end. A more reasonable choice would be the envelope of the Case B curve and of the curve of present U.S. reasonably assured and potential uranium resources. We call such a schedule Case B'. If the schedule of Case B is adopted, the net discounted benefit of LMFBR is reduced from $14.7 billion for the base case to $7.3 billion. Using Case B', the benefit is reduced to $6.8 billion. The supply-versus-price schedules for cases B and B' are shown in Table 7.

One feature that distinguishes uranium ore from other minerals or energy resources is its radioactivity. Consequently, uranium deposits at or near the surface can be easily detected, and these deposits at present constitute most of our low-cost reserves. This implies that

Table 7

PROJECTED URANIUM SUPPLY-VERSUS-PRICE SCHEDULES

Cumulative Supply (million tons U_3O_8)	Case B ($/lb)	Case B' ($/lb)
0	8	8
1	10	10
2	14	14
3	18	18
4	25	25
5	40	39
6	50	42
7	60	45
8	70	49
9	75	54
10	80	59

Source: Case B from USAEC, *Proposed Final Environmental Statement*, WASH-1535, p. 11.2-72, and Case B' from author.

the chance of finding additional low-cost uranium ore may be less than that of finding other types of deposits at the corresponding stage of reserve development. Since Case B′ has only been modified near the high price end ($\geq$ $40 per pound), the above concern should be of little consequence in our case. There was also a concern in CB-70 that even the supply in Case B might not be realized: "If the breeder is introduced in the 1980's as expected, the uranium demand can be expected to peak during the next thirty years. If this prediction curtails prospecting efforts, this projection may not be realized." [4] This may well be true. But it is of no concern to us because we are calculating the difference in energy costs with and without the LMFBR in the electric power system. To compute energy costs without the LMFBR, one should, in principle, use a supply curve resulting from no LMFBR introduction. Another supply curve, lower than the former because of less profit incentive for exploration, should be used to compute energy costs for the electric power system with the LMFBR. If one uses the supply curve with expected LMFBR introduction for both situations, as was done in CB-74, the LMFBR benefit is inflated.

---

[4] USAEC, *Updated (1970) Cost-Benefit Analysis*, WASH-1184, p. 43.

4

PLANT CAPACITIES, ELECTRICAL ENERGY DEMAND, AND PLANT MIX

Overrestriction of High-Temperature Gas Reactor Plant Capacity

In CB-74, the plant capacities of HTGR and LMFBR are restricted in the base case. The rationale is stated as follows:

> Without proper constraints, the linear program might choose to build new plants at a rate which exceeds industry growth capacities. Thus, introductory constraints are required for LMFBRs and HTGRs to simulate the tooling-up of the nuclear industry for production of new plant types. For those cases in which introduction constraints are limiting, they were applied as follows: Two HTGRs, some 2000 MWe are currently committed for operation in the 1980–81 biennium. In the following biennium, 4000 MWe maximum were allowed; in following periods, the HTGR capacity could not exceed twice the capacity of the preceding biennium. Only one LMFBR is permitted in the year of introduction. With several potential vendors, a maximum of 8000 MWe of LMFBR capacity was permitted in the next biennium, with the maximum new capacity in each following biennium limited to twice the capacity of the preceding biennium. In the years past 2000, breeder and nonbreeder plants compete within the assumed nuclear envelope. The HTGR penetration was limited to 25 percent of the nonbreeder portion of this envelope in acknowledgement of the established competitive position of LWRs.[1]

With essentially the same plant capital cost as LWRs but half the uranium consumption and higher efficiency, HTGRs will play a

---

[1] USAEC, *Proposed Final Environmental Statement*, WASH-1535, pp. 11.2-63 to 64.

Figure 10

ANNUAL RATES OF NUCLEAR CAPACITY ADDITIONS WITH LMFBR

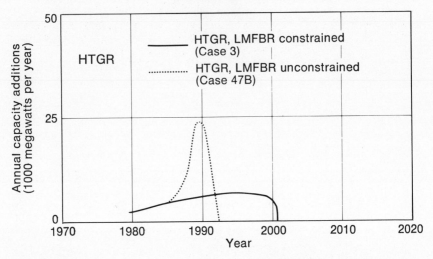

Source: USAEC, *Proposed Final Environmental Statement*, WASH-1535, vol. 4, p. 11.2-114.

major role in the nuclear power system without the LMFBR. In the case with LMFBR, the constraint does not affect the HTGR capacity appreciably. This is shown in Figure 10.

The constraints on HTGRs adopted by the AEC for its linear program certainly have put the benefits of the LMFBR under a very favorable light. The 40-MWe prototype HTGR was placed in commercial operation in May 1967 and the 330-MWe HTGR is expected to be in commercial operation soon. The utilities have already placed several orders for 770-MWe and 1160-MWe plants, some of which will be in commercial operation by 1980–81. Yet, HTGR capacity is restricted to an artificially low level. Even after the year 2000, twenty years after the introduction of the large-scale HTGR, this reactor is still restricted to no more than 25 percent of the non-breeder nuclear capacity. In contrast, the constraint on LMFBR capacity is to be completely removed thirteen years after the LMFBR's commercial introduction. The AEC argues that:

> The capital investment required for a second vendor to enter the HTGR Nuclear Steam Supply (NSSS) market would be in the range of hundreds of millions of dollars, perhaps as much as $500 million. In the 1980's, the present LWR NSSS

manufacturers probably would be starting to get a good return on their investment and if the utilities did not indicate a strong preference for the HTGR, they might be reluctant to make the investment required to be able to offer HTGRs.[2]

Given the HTGR's distinct advantage of lower fuel cost and the fact that it has no definite disadvantages in plant capital cost and environmental hazards, what would prevent the utilities from choosing HTGRs? Even if they should reject HTGRs on noneconomic and nonenvironmental grounds, a cost-benefit analysis should still assume an electric generating system that minimized the cost of electricity. When the demand for HTGRs soars, more manufacturers will enter into the field. The AEC also presented the argument that "[t]he HTGR penetration to the year 2000 in the nonbreeder cases is in good agreement with the vendor's (GAC) [General Atomic Corporation] own estimate."[3] Underestimation of HTGR capacity will discourage the entry of competitors and is advantageous to the sole vendor at the present time.

Plant-Capacity Constraints Used in the Present Study

There is, however, some justification for restricting the initial capacity of HTGRs and LMFBRs during the tooling-up period because a new product involves new technology in construction and mass production. But it is not equitable to remove the constraint on LMFBR capacity after the year 2000 while restricting HTGR capacity beyond the year 2000 to less than 25 percent of the nonbreeder capacity. Since the AEC proposed that the constraint on LMFBR capacity be removed thirteen years after the LMFBR's commercial introduction, the constraint on the HTGR should also be removed thirteen years after the introduction of large scale HTGRs, that is, in 1993. As shown in Figure 11, for the HTGR restricted and unrestricted cases, the differences in annual rates of capacity additions of HTGRs and LWRs are not appreciable until 1993, but the differences are substantial thereafter. A perturbative calculation thus suffices and shows that the net discounted benefit of LMFBRs, with constraints on HTGRs and LMFBRs for the initial thirteen years and a Case B uranium supply, is reduced from $7.3 billion to $4.1 billion (Figure 7). With a Case B' uranium supply, the benefit is reduced from $6.8 billion to $3.7 billion. Results for the same cases but with unrestricted HTGR and LMFBR capacities are also shown in Figure 7 for comparison.

---

[2] Ibid., p. 11.2-65.
[3] Ibid.

Figure 11

ANNUAL RATES OF NUCLEAR CAPACITY ADDITIONS WITHOUT LMFBR

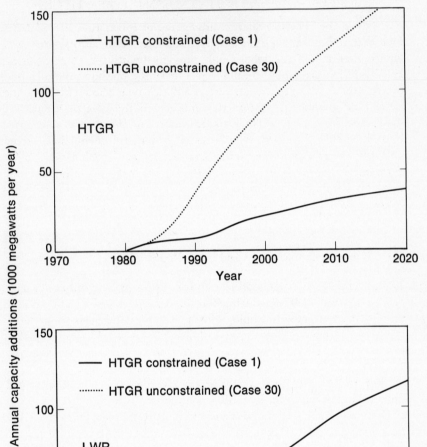

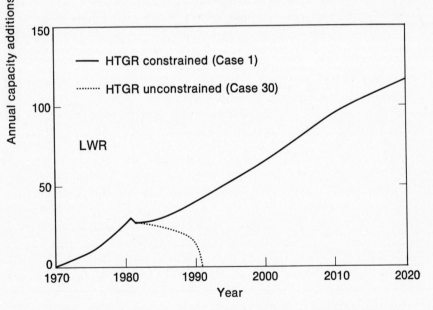

Source: USAEC, *Proposed Final Environmental Statement,* WASH-1535, vol. 4, p. 11.2-112.

Table 8
ENERGY PROJECTIONS USED IN CB-74

	1970	1980	1985	2000	2020
Population (millions)	205	228	240	271	307
GNP (constant 1958 billions of dollars)	724	1,141	1,360	2,300	4,291
Total energy demand (Btu $\times$ 10^{15})	67	99	117	195	359
Gross energy inputs to electric utilities (Btu $\times$ 10^{15})	17	30	40	97	235
Gross inputs to utilities as percent of total energy	25	30	34	50	65
Electric generating capacity (in 1000 MWe)					
Nuclear			260	1,200	3,300
Hydro			97	150	210
Internal combustion and combustion turbine			71	150	250
Fossil and other			486	700	2,130
Total			914	2,200	5,890

Source: USAEC, *Proposed Final Environmental Statement*, WASH-1535, vol. 4, p. 11.2-53.

Projections of Electrical Energy Demand

The projections of total and electrical energy demand used in CB-74 (see Table 8) follow the approach used by the AEC's Office of Planning and Analysis [4] and are in agreement with its Case B projection, which is described as follows:

> Case B represents a projection of energy growth which assumes a continuation of the past relationship between energy consumption and GNP, together with a further increase in the importance of electricity as a secondary energy source. On this basis total energy consumption was projected to increase at a 3.6 percent annual compound rate. Thus, while per capita energy use is expected to grow, this projection also implies that the declining long-term trend

[4] U.S. Atomic Energy Commission, Forecasting Branch, Office of Planning and Analysis, *Nuclear Power Growth 1974-2000*, WASH-1139(74) (February 1974), p. 12.

in energy input required to produce a dollar of output will continue. Furthermore, the projection indicates that the share of total primary energy required by the year 2000 for electricity generation will nearly double as it has for a similar time period in the past.

Case B inherently assumes that factors historically important in shifting the pattern of energy use in favor of electricity will influence future demand. Electricity is expected to remain a useful, convenient, and inexpensive form of energy relative to available substitutes.[5]

These projections are based mainly on historic patterns of energy consumption and the electric share of energy consumption, at a time when energy was traditionally cheap. But recently the prices of energy, that of electricity in particular, have risen sharply and to a greater extent than the general price level. If this continues, electricity consumption and total energy consumption will fall well below the predictions used in CB-74, for that report does not consider the effect of price elasticity. The conclusion of Chapman, Tyrell and Mount is that the annual electric consumption rate in the year 2000 will probably be in the range of 1.9 to 4.6 trillion kWhr if the elasticity factor for electricity price, as well as various assumptions about population growth and the increase in electric cost, is used.[6] The annual electric consumption rate in the year 2000 used in CB-74 is 10.6 trillion kWhr, which is substantially higher than the above projection. A 20 percent reduction in energy demand by the year 2020 reduces the net discounted benefit of the LMFBR from $14.7 billion for the AEC's base case (Case 3 in CB-74) to $9.5 billion (Case 15). A 50 percent reduction in energy demand results in a benefit of only $2.2 billion. For our cases shown in Figure 7, which assume the Case B or B' uranium supply schedule and further that the HTGR and LMFBR are restricted for thirteen years or unrestricted, a 50 percent reduction in energy demand by the year 2020 will wipe out all the net discounted benefit of the LMFBR.

Energy demand is projected by other agencies and individuals. Those mentioned in CB-74 are summarized in Table 9.

It is apparent that the demand projections in CB-74 are near the high end of the estimates. Lower electric energy demand would, of course, result in lower LMFBR benefits.

---

[5] Ibid., p. 8.

[6] Diane Chapman, Timothy Tyrell and Timothy Mount, "Electricity Demand Growth and the Energy Crisis," *Science*, vol. 178 (1972), pp. 703-8, quoted in USAEC, *Proposed Final Environmental Statement*, WASH-1535, vol. 4, p. 11.2-56.

Table 9

PROJECTED ENERGY DEMAND IN YEAR 2000

Authors	Total Energy (Btu $\times$ 10^{15})	Electric Energy (10^{12} kWhr)
AEC (CB-74)	195	10.6
AEC (CB-70)		10.0
Ford Foundation Energy Report	185	
Survey of Energy Consumption Projections	135-213	4.5-10.8
Bureau of Mines	166-239	
Milton F. Searl		5.5
Chapman, Tyrell and Mount		1.9-4.6
Federal Power Commission		6.1-12.3
Department of the Interior		9.0

Source: USAEC, *Proposed Final Environmental Statement*, WASH-1535, vol. 4, p. 11.2-53; USAEC, *Updated (1970) Cost-Benefit Analysis*, WASH-1184, p. 45; Energy Policy Project of the Ford Foundation, *Exploring Energy Choices* (Washington, D. C., 1974); *Survey of Energy Consumption Projections*, prepared at the request of Senator Henry M. Jackson, chairman, Senate Committee on Interior and Insular Affairs (Washington, D. C.: U.S. Government Printing Office, 1972); Milton F. Searl quoted in Cochran, *Fast Breeder Reactor*, p. 110; Chapman et al., "Electricity Demand Growth," pp. 703-8; Federal Power Commission, *Oil and Gas Input Issues*, Hearings before the Senate Committee on Interior and Insular Affairs, January 10, 11, and 22, 1973, p. 629; and Walter G. Dupree, Jr. and James A. West, "United States Energy through the Year 2000," U.S. Department of the Interior, December 1972.

There are other independent projections which are summarized and reviewed by Cochran.[7] In conclusion, Cochran stated that:

> Current long-range electrical energy demand projections, using independent forecasting techniques that are based on historical national trends in GNP growth, income, gas and electricity price elasticities, and per capita consumption, suggest that the 1970 Survey, and in turn the 1970 Analysis [CB-70], projections overestimate the electrical demand. The true demand could easily be 25 percent, and possibly 50 percent, below the "probable" projection in the 1970 Analysis for the year 2000. If these projections are more correct than the 1970 Survey, the projected discounted net benefits of the LMFBR program—without changing the remaining eco-

[7] Thomas B. Cochran, *The Liquid Metal Fast Breeder Reactor, An Environmental and Economic Critique* (Washington, D. C.: Resources for the Future, Inc., 1974), pp. 110-15.

nomic and technologic projections from the most probable estimates in the 1970 Analysis—could vanish, due only to the reduction in energy demand.[8]

With the understanding that the projections used in CB-74 are probably too high and the LMFBR benefits overestimated, we adopt these projections so as to ensure that the LMFBR is most favorably viewed in this regard.

The Role of Coal-Fired Plants in the Future Electric Power System

With a given electrical energy demand, the rate of annual additions of LMFBR is still dependent on the competitiveness among nuclear, fossil-fueled, and other types of power plants. In Table 8, the two largest components of the plant mix as determined in CB-74 are nuclear and fossil. In CB-68, the fossil-fueled capacity installed after 1989 was projected at 13 percent of the nuclear installations because of the lower future fossil fuel costs, then projected. It was zero percent in CB-70.[9] In CB-74, the fossil capacity installed after 1985 is 54 percent of the nuclear installations.[10] Thus, it is apparent that the plant mix is very sensitive to the assumed plant and fuel costs of nuclear and fossil-fueled plants. Fossil-fueled power plants are fueled by gas, oil, and coal. Since the electric energy supplied by the gas- and oil-fueled power plants is expected to decline in the future, these power plants do not play a significant role in our cost-benefit analysis.

The competitiveness of coal-fired plants is sensitive to the cost of coal. The sizable share of fossil plants in CB-68 was due to the projected gradual decline of coal prices to about $6 per ton until 1978 followed by an increase of only 0.1 percent per year. Occurrences since 1968, such as more stringent mine safety regulations, tighter environmental standards, and higher labor costs may have broken the historic downward trend in coal prices. In CB-70, the projected price was raised to $8 per ton in 1972 and beyond. Such a slight price increase has shown its significant effect in reducing the share of fossil plants in the new installation market after 1989 to zero. The coal price has roughly doubled since 1970 even after adjustment for inflation. In addition, the installation of sulfur-oxide removal systems to comply with the forthcoming new air-quality standards will in-

---

[8] Ibid., pp. 116-17.

[9] USAEC, *Updated (1970) Cost-Benefit Analysis*, WASH-1184, p. 34.

[10] USAEC, *Proposed Final Environmental Statement*, WASH-1535, vol. 4, p. 11.2-58.

Table 10

LWR PLANT CAPITAL COST TREND

(in 1974 dollars)

Date of Estimation	Cost per kWe Capacity
March 1967	179
June 1969	259
January 1971	339
January 1971 (revised)	393
January 1973	461
February 1973	475
December 1973	436-500

Source: Irvin C. Bupp and Jean-Claude Derian, "Another Look at the Economics of Breeder Reactors" (Cambridge: Center for International Affairs, Harvard University, November 1973), p. 12; and USAEC, *Nuclear Industry*, WASH-1174(73), p. 12.

crease the capital plant cost of coal-fired power plants. However, there has also been an alarming rise in the construction costs of nuclear power plants, as shown in Table 10. The plant capital costs include direct costs, such as land, structures and site facilities, reactor, turbine, electric plant equipment, contingencies, and indirect costs, such as professional services and interest during construction. When one combines all the factors, the electric generating cost of a coal-fired plant in operation by 1981 is only somewhat higher than that of an LWR, as is shown in Table 11. The coal price is, however,

Table 11

ESTIMATED GENERATION COST[a] FOR 1,000 MWe STEAM-ELECTRIC POWER PLANTS IN 1981

(mills/kWhr, in 1973 dollars)

Cost Component	LWR	Coal	Oil
Capital	11.70	10.90	8.00
Fuel	2.50	5.50	24.60
Operation and maintenance	1.00	1.60	0.80
	15.20	18.00	33.40

[a] All costs were escalated at 5 percent per year with the exception of nuclear fuel costs.

Source: USAEC, *Nuclear Industry*, WASH-1174(73), p. 15.

highly dependent on geographic location. Approximately half of the average coal price on an "as burned" basis is attributable to transportation cost. In CB-70, the estimated future coal prices in different districts of the country varied from $4.96 per ton to $10.39 per ton, with an average of $8 per ton. In CB-74, the average cost of coal to U.S. utilities is projected to increase from $15.30 in 1974 to $19.13 (in 1974 dollars) in 2020.[11] Even if an average coal-fired plant may not be competitive against an LWR, plants built in low coal-cost districts may still be able to generate electricity at an attractive cost. Also shown in Table 11 is the generation cost of an oil-fueled power plant. Its extremely high fuel cost will be likely to rule it out of the installation market in the period of our consideration, which is 1980 to 2020.

With coal resources alone sufficient to meet U.S. energy requirements for hundreds of years and with the competitiveness of coal against nuclear power plants at least in low coal-cost districts, the chance is good that coal-fueled power plants will maintain a sizable share of the new installation market within the period of our consideration. However, we consider the market share of fossil plants used in CB-74 and shown in Table 8 to be reasonable.

Nuclear Power Forecasts by the Atomic Energy Commission

Since the projections of total electricity demand and the fossil-fueled capacity additions contained in CB-74 are used in the present study, the projection of nuclear capacity additions contained in CB-74 also has to be accepted for reason of consistency. However, one may still be interested in comparing this projection with other AEC projections on nuclear capacity, which are shown in Table 12. The projections up to the year 2000 do not fluctuate appreciably. However, the nuclear capacity beyond 2000 used in CB-74 has been revised downward from CB-70.

In conclusion, since we do not alter the energy demand and plant mix used in CB-74, the net discounted benefits of the LMFBR for the various cases considered in our present study remain the same as those quoted at the end of the previous chapter. The unchanged mix also explains the fact that the dollar figures emerging from the block of energy demand and plant mix in Figure 7 are the same as those entering it.

---

[11] Ibid., p. 11.2-101.

Table 12
AEC NUCLEAR POWER FORECASTS

Year Forecast Made	Total Capacity at End of Calendar Year, in 1,000 MWe				
	1975	1980	1985	2000	2020
1962	16	40	100	734	
1964	29	75			
1966	40	95			
1967	61	145	255		
1970	59	150	300		
1970 (CB-70) [a]	60	140	280	1,380	4,000
1971	57	151	306		
1972 (most likely case)	54	132	280	1,200	
1974 (Case B)	47	102	260	1,200	
1974 (CB-74)	54	132	260	1,200	3,300

[a] Projection used in CB-70 for the base case (Case 3). Interpolation of data has been employed.

Source: USAEC, *Nuclear Power Growth,* WASH-1139(74), p. 17; USAEC, *Updated (1970) Cost-Benefit Analysis,* WASH-1184, p. 34; and USAEC, *Proposed Final Environmental Statement,* WASH-1535, vol. 4, p. 11.2-113.

5

COST PROJECTIONS BY THE ATOMIC ENERGY COMMISSION

Fuel-Cycle Costs of Light-Water Reactors

A comparison of the costs of generating electricity by LWRs for three periods is shown in Table 13.

Operating and maintenance costs plus fuel-cycle costs for LWR plants in operation by 1981, as estimated in 1973, are not significantly different from those for plants in operation by 1970 and 1978. The plant capital cost, however, changed substantially over the years. We now take a closer look at the components of fuel-cycle cost for plants to be in operation by 1981, as shown in Table 14. In CB-74 the fuel generation cost of an LWR is not broken down in as much detail as it is in Table 13. However, the levelized fuel cost as shown in Table 15 gives a good indication of the fuel costs used in CB-74.

Table 13
ELECTRIC GENERATING COST TREND OF LWR
(mills/kWhr, in 1974 dollars)

Year of Estimation	1967	1971	1973
Year of Plant in Operation	1970	1978	1981
Capital cost	3.27	7.99	10.95
Operation and maintenance cost	1.63	0.52	0.94
Fuel-cycle cost	2.07	1.76	2.34
Total cost of electricity	6.97	10.27	14.23

Source: USAEC, *Nuclear Power Growth Pattern*, WASH-1098 (December 1970); USAEC, *Nuclear Industry*, WASH-1174(71), p. 91; USAEC, *Nuclear Industry*, WASH-1174(73), p. 15.

Table 14

ESTIMATED 1981 FUEL-CYCLE COSTS
OF LWR FOR 1,000 MWe PLANT
(mills/kWhr, in 1974 dollars)

Cost Component	Cost
Mining and milling ($10.9/lb U_3O_8)	0.59
Conversion to UF_6 ($1.47/lb U)	0.08
Enrichment ($46/kg SWU)	0.83
Reconversion and fabrication ($76/kg U)	0.36
Spent fuel shipping ($5.4/kg U)	0.02
Reprocessing ($38/kg U)	0.15
Waste management	0.04
Plutonium credit ($7.0/g)	(0.24)
Subtotal	1.83
Fuel inventory carrying charge (at 12%)	0.89
Total	2.72

Source: USAEC, *Nuclear Industry*, WASH-1174(73), p. 15.

Table 15

THIRTY-YEAR LEVELIZED FUEL-CYCLE COSTS
FOR PLANTS BUILT IN 1990-91
(in 1990 dollar values)

Plant Type	Case 1 (No Breeder) Plants added (1000 MWe)	Case 1 (No Breeder) Fuel cost (mills/kWhr)	Case 3 (1987 LMFBR) Plants added (1000 MWe)	Case 3 (1987 LMFBR) Fuel cost (mills/kWhr)
LWR (U fueled)	61	3.93		
LWR (U-Pu fueled)	6	3.97	65	3.55
LWR (Pu fueled)	17	3.49		
HTGR	19	3.65	11	3.19
LMFBR (advanced oxide)			26	1.96

Source: USAEC, *Proposed Final Environmental Statement*, WASH-1535, vol. 4, p. 11.2-192.

The higher fuel-cycle cost for the LWR in Table 15, compared with that for the 1981 plant shown in Table 14, is largely due to the higher cost of U_3O_8 assumed in the calculations of Table 15. However, the effect of uranium cost has been accounted for in Chapter 3.

The fabrication cost trend used in CB-74 is consistent with the estimated cost of $76 per kgU by 1981. But, the reprocessing cost in CB-74 is higher. The enrichment cost is a constant $36/kgSWU in CB-74 compared with $46/kgSWU for the 1981 plant. On February 4, 1973, the AEC announced that, effective August 14, 1973, the charge for enrichment services would be $38.50/kg instead of $32/kg. Also, the unit charges would be increased automatically by 1 percent every six months beginning January 1, 1974. However, the unit charge was raised to $44.25/kg effective December 1974, which is higher than the charge according to the proposed schedule.

One should not overlook the feasibility of laser enrichment. In October 1973, Exxon Nuclear president, Raymond L. Dickeman, told the congressional Joint Committee on Atomic Energy that "a commercial laser enrichment process could be operating by the mid-1980's at an overall cost of 10 to 20 percent less than the cost of gas centrifuge techniques."[1] In January 1974, the AEC's general manager, John A. Erlewine, discussed the implications of the above development in a letter to the joint committee. His statement was quoted in *Science* as follows:

> If laser techniques lived up to their current promise of low cost and high efficiency, Erlewine said, such a process would "make alternative enrichment processes economically obsolete."
>
> There were implications for the nation's breeder reactor program as well, Erlewine acknowledged. The AEC has predicted its argument for pressing rapidly ahead with the breeder—which would make plutonium fuel—on a prediction that the present low cost of uranium fuel will begin to soar in the mid-1980's as reserves of high grade ore diminish. Commercial laser enrichment, Erlewine said, could reduce natural uranium demand by 10 to 40 percent and "establish a more difficult economic target for the commercialization of the breeder."[2]

The future conversion cost estimated in 1969 was $1.00/lb U or $1.31/lb U in 1974 dollars,[3] compared with the $1.47/lb U shown in Table 14. The difference of $0.16/lb U amounts to only 0.009 mill/kWhr, which can be ignored within the accuracy of our estimates.

---

[1] Robert Gillette, "Uranium Enrichment: Rumors of Israeli Progress with Lasers," *Science*, vol. 183 (March 22, 1974), p. 1173.

[2] Ibid.

[3] R. E. Aronstein et al., *1000 MWe Liquid Metal Fast Breeder Reactor Follow-On Study Conceptual Design Reprint*, AI-AEC-12792 (1969), vol. 2, p. 253.

The value of fissile plutonium for recycling is dependent on the LWR-HTGR-LMFBR mix in the electric power system. Any change in plutonium value is internally corrected and accounted for in the analysis. The variations in estimating the costs of spent fuel shipping and waste management are too small to warrant detailed examination here. Therefore, the components of the fuel-cycle cost of the LWR used in CB-74 are either accounted for in the present study or acceptable to us. The same conclusion holds for the components of the fuel-cycle costs of the HTGR and the LMFBR.

Reactor Characteristics of Nuclear Reactors

The reactor characteristics of the LWR and the HTGR are assumed in this study to be the same as those used in CB-74, which are shown in Table 16. Should any characteristic be modified, all relevant factors should be considered. For example, higher irradiation levels would decrease fabrication and processing costs per unit of energy produced.[4] But higher burnup would increase the quantity of fission gas released during operation, and requires thicker cladding. Thicker cladding causes higher neutron absorption, thus requiring fuel with higher enrichment. The fixed charges on fuel will also increase because of larger inventory. In addition, higher enrichment is needed to compensate for the increase in fission product poisons resulting from higher irradiation.

The reactor characteristics of the LMFBR are also assumed to be the same as those used in CB-74, which are shown in Table 16. This, however, does not imply that the AEC has conservatively determined the performance characteristics of the LMFBR. In fact, in CB-70, the early oxide LMFBR was assumed to have a core burnup of 68,000 Mwd/tonne compared with the much more optimistic burnup level of 110,000 Mwd/tonne assumed in CB-74. Also, as pointed out by Cochran:

> The initial specific inventory of the *early* oxide LMFBR in the 1973 Analysis [CB-74D], 1.88 kg fissile/MWe, represents a considerable improvement over the initial specific inventory of the *advanced* oxide LMFBR in the 1970 Analysis [CB-70].[5]

---

[4] Samuel Glasstone and Alexander Sesonske, *Nuclear Reactor Engineering* (Princeton, N. J.: Van Nostrand Co., 1963), p. 794.

[5] Thomas B. Cochran and Arthur R. Tamplin, "NRDC Comments on WASH 1535," quoted in USAEC, *Proposed Final Environmental Statement*, WASH-1535, vol. 6, p. 38-222.

Table 16

TYPICAL REACTOR CHARACTERISTICS ASSUMED IN CB-74

Reactor Design	Plant Net Thermal Efficiency (percent)	Earliest Date of Availability	Fuel	Equilibrium Fuel Exposures (MWD/tonne)	Specific Power (MW/tonne)	kg/MWe-Year Net yield fissile plutonium	kg/MWe-Year U-233	kg/MWe-Year Net consumption U-235	Plutonium Simple Doubling Time (years)	Net U_3O_8 (tonnes/MWe-year)	Net Separative Work (SWU/MWe-year)
LWR											
Nonrecycle	33.0	1970	Enriched uranium for 30 years	32,000	37.3	0.228		0.851		0.211	148
Full plutonium recycle	33.0	1986	Fully enriched plutonium in depleted uranium for 30 years	32,000	37.3	−0.377		0.062		0[a]	0
Intensive plutonium	33.0	1980	Enriched uranium for 30 years with intensive plutonium production configuration	15,000	23.6	0.460		1.016		0.253	179
HTGR											
Annual refueling	38.5	1980	Highly enriched U-235 startup, U-233 recycle	81,500	76.6	0.001	0.050	0.404		0.110	103
Semiannual refueling	38.5	1984	Highly enriched U-235 startup, U-233 recycle	66,000	69.0	0.001	0.050	0.370		0.101	95
Plutonium recycle	38.5	1986	Plutonium fuel in thorium carbide	199,000	249.2	−0.715	0.125	−0.003		0	0

Table 16 (*continued*)

| Reactor Design | Plant Net Thermal Efficiency (percent) | Earliest Date of Availability | Fuel | Equilibrium Fuel Exposures (MWD/tonne) | Specific Power (MW/tonne) | Net yield fissile plutonium | kg/MWe-Year | | Plutonium Simple Doubling Time (years) | Net U$_3$O$_8$ (tonnes/ MWe-year) | Net Separative Work (SWU/ MWe-year) |
							U-233	Net con-sumption U-235			
LMFBR											
Early oxide	40.0	1987	Plutonium-fueled oxide	110,000-core 6,000-blanket	53.0	0.141		0.011	28.0	0[a]	0
Advanced oxide	40.0	1991	Plutonium-fueled oxide	110,000-core 6,000-blanket	72.0	0.247		0.010	12.5	0[a]	
Advanced carbide (small diam pin)	40.0	1995	Plutonium-fueled oxide	110,000-core 6,000-blanket	60.0	0.303		0.013	7.6	0[a]	
Advanced carbide (large diam pin)	40.0	1995	Plutonium-fueled carbide	110,000-core 6,000-blanket	46.9	0.381		0.012	7.4	0[a]	
Advanced uranium oxide	40.0	1991	Uranium-fueled oxide	110,000-core 4,500-blanket	72.0	0.507		1.637		0.372	327
Advanced uranium carbide	40.0	1995	Uranium-fueled carbide	95,000-core 8,600-blanket	60.0	0.598		1.629		0.359	312

a Uses uranium from tails stockpile.

Source: USAEC, *Proposed Final Environmental Statement,* WASH-1535, vol. 4, p. 11.2-68.

Cochran's conclusion is as follows:

> In other words, the characteristics of advanced oxide LMFBR in the 1970 Analysis [CB-70] was based on a design that optimized performance at the expense of safety considerations, could not be licensed today, and has been shown since to be infeasible due to the stainless steel swelling phenomena. Nevertheless, in the 1973 Analysis [CB-74D] the oxide-fueled LMFBR have even superior performance characteristics, but there are no detailed LMFBR designs (at least none available to the public or the AEC) from which these characteristics are derived. It would be interesting to know the safety characteristics of LMFBRs having these performances. None are given. We seriously question whether these specific inventories are representative of anything close to realistic designs.[6]

In this study, we assume that the performance characteristics of the LMFBR as described in CB-74 can be achieved so as not to underestimate the benefit of the LMFBR.

Plant Capital-Cost Differential

The plant capital cost of a nuclear reactor has risen substantially. As shown in Table 10, the cost of an LWR plant has more than doubled since 1967. The cost of the 1300 MWe Midland nuclear plant of Consumers Power Co., estimated at $268/kWe in October 1968 (in 1968 dollars),[7] had soared to $1080/kWe (in 1975 dollars) by April 1975. Though the Midland plant had more than its fair share of problems, many of its cost-overrun factors are common to other nuclear plants under construction.

The key factor is, however, the capital-cost differential between a thermal reactor (LWR and HTGR) and an LMFBR. The capital cost of an LMFBR is expected to increase correspondingly. The capital costs of fossil-fueled and nuclear power plants used in CB-68 and CB-70 analyses are shown in Figure 12, while those used in CB-74 are shown in Figure 13.

In CB-68, the capital-cost differential between an LMFBR, commercially introduced, and an LWR was estimated to be around $35/kWe, which is about 25 percent of the then estimated LWR plant cost. In CB-70, the differential between an LMFBR and an LWR

---

[6] Ibid., p. 38-228.

[7] Fred E. Garrett, "What a Price to Pay, Nuclear Plant Earns a Dubious Distinction," *Saginaw News*, April 20, 1975, Section D, p. 1.

Figure 12

PLANT CAPITAL COSTS USED IN CB-68 AND CB-70

(in 1970 dollars)

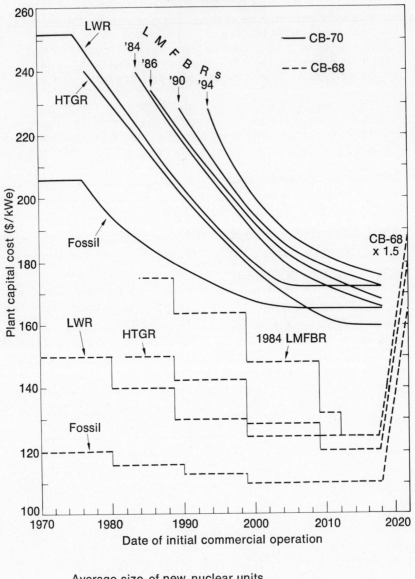

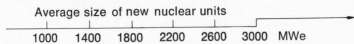

Average size of new nuclear units

Source: USAEC, *Updated (1970) Cost-Benefit Analysis*, WASH-1184, p. 37.

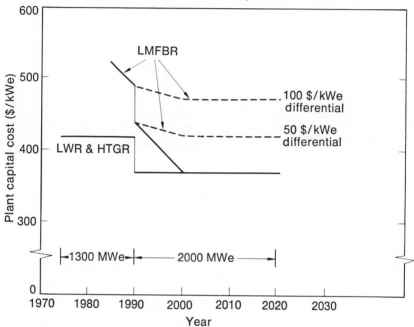

Figure 13

PLANT CAPITAL COSTS USED IN CB-74

(in 1974 dollars)

Source: USAEC, *Proposed Final Environmental Statement*, WASH-1535, vol. 4, p. 11.2-12.

(or HTGR) was reduced to around $21/kWe, which amounts to only about 10 percent of the LWR (or HTGR) plant cost. In CB-74, the cost differential was $67/kWe or about 16 percent. Even though the major difference in cost is from the reactor plant equipment, which accounts for about 30 percent of the total construction cost, a 16 percent higher capital cost does not allow sufficient room for unexpected cost overruns, which is quite common for a new technology. A much more questionable assumption used in CB-74 was to allow the cost differential to drop to zero beyond the year 2000. In CB-68 and CB-70, a finite difference was assumed at least until the year 2010. Also, no learning curve was assumed for the LWR, the HTGR, and coal-fired fossil plants while an extremely optimistic learning curve was assumed for the LMFBR. Cochran summarized the sentiment as follows:

> The Capital Cost Evaluation Group at Oak Ridge National Laboratory (ORNL) has studied nuclear (and fossil) plant

Figure 14

LEARNING FACTORS OF PLANT CAPITAL COSTS
BY OAK RIDGE NATIONAL LABORATORY

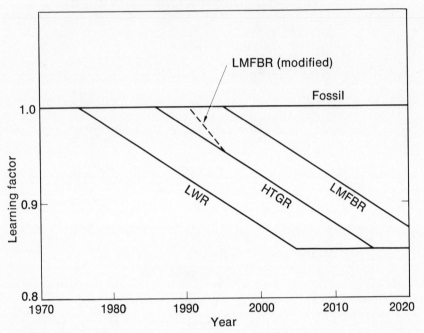

Source: Howard I. Bowers and M. L. Myers, "Estimated Capital Costs of Nuclear and Fossil Power Plants," ORNL-TM-3243 (Oak Ridge National Laboratory, March 5, 1971), p. 20, quoted in Thomas B. Cochran and Arthur R. Tamplin, "NRDC Comments on WASH 1535," in USAEC, *Proposed Final Environmental Statement*, WASH-1535, vol. 6, p. 38-179.

costs extensively for a number of years. The AEC relies principally on the ORNL staff to provide cost data for the cost-benefit analyses. It is significant to note that ORNL staff proposed at the time of the 1970 Analysis [CB-70] that all nuclear plants should use a 95% learning curve (5% improvement per decade) starting 10 years following commercial introduction. These ORNL learning curves are shown in Figure 2(a) [Figure 14]. In the 1970 Analysis [CB-70], the AEC did not follow ORNL's suggestion. Instead, AEC used the modified learning curve shown in Figure 2(a) [Figure 14]. As noted in the 1970 Critique the modified LMFBR learning curve reduced the LMFBR-LWR capital cost difference more rapidly and gave the LMFBR a capital cost advantage which it would not have had under

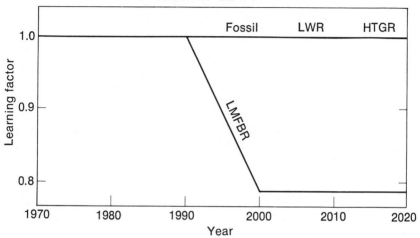

Figure 15

LEARNING FACTORS OF PLANT CAPITAL COSTS
USED IN CB-74

Source: Cochran and Tamplin, "NRDC Comments," in USAEC, *Proposed Final Environmental Statement*, WASH-1535, vol. 6, p. 38-179.

ORNL assumptions. The learning curves used in the 1973 Analysis [CB-74D], plotted in Figure 2(b) [Figure 15], contrast sharply with those from the previous analysis [Figure 2(a)].[8]

Bupp and Derian gave a similar observation:

Breeders will not be competitive with LWRs when only 10-20 of the former have been installed as opposed to several hundred of the latter. Consequently, an added cost will be incurred by breeders until their manufacturing costs have decreased enough to compensate for the initial LWR advantage due to fabrication experience. These learning-curve costs must be taken into account even if it is assumed that breeder capital costs will turn out to be below the allowable threshold for previous U_3O_8 prices and other variables. In such circumstances, learning curve costs have to be subtracted from the breeder advantage in order to determine the real benefits of their introduction.[9]

---

[8] Cochran and Tamplin, "NRDC Comments," in USAEC, *Proposed Final Environmental Statement*, WASH-1535, vol. 6, p. 38-178.

[9] Irvin C. Bupp and Jean-Claudé Derian, "Another Look at the Economics of Breeder Reactors" (Cambridge: Center for International Affairs, Harvard University, November 1973), p. 34.

Table 17
ESTIMATED LMFBR PROGRAM COSTS
FROM MID-1974 TO 2020
(billions of mid-1974 dollars)

	LMFBR Introduction Dates		
	1985	1987	1991
AEC costs			
LMFBR	6.6	6.5	6.8
Supporting technology	1.6	1.6	1.6
Electric utility contribution to CRBR[a]	0.26	0.26	0.26
Total	8.6	8.4	8.7
10% discount to 1974	4.8	4.7	4.7

[a] Clinch River Breeder Reactor demonstration plant (CRBR).
Source: USAEC, *Proposed Final Environmental Statement*, WASH-1535, vol. 4, p. 11.2-33.

In Chapter 4, we discussed the HTGR and LMFBR introduction rate and its restriction during the initial years. A slower introduction rate causes LMFBR cost reduction to decline even more slowly. In my view, a $50/kWe cost differential as shown in Figure 13 should be maintained between the LMFBR and the LWR (HTGR) to approximate the learning effects. This cost differential turns the benefits of the LMFBR into losses for all cases under consideration in the present study (Figure 7).

Costs of the Liquid Metal Fast Breeder Reactor Program

The LMFBR program costs used in CB-74 are shown in Table 17. The program costs include only $260 million contributed by the electric utilities for the Clinch River Breeder Reactor (CRBR) demonstration plant. The $18 billion to be committed by the utilities through 1984 is not included,[10] since these costs are assumed to be recoverable through the sale of nuclear plants, fuels, and services. The breakdown of the LMFBR program cost is shown in Table 18.

[10] *AEC Authorizing Legislation—Fiscal Year 1972*, Hearings before the Joint Committee on Atomic Energy, 92nd Congress, 1st session, 1971, p. 96.

Table 18

DETAIL OF LMFBR PROGRAM COST PROJECTIONS (1975 THROUGH 2020), LMFBR INTRODUCTION DATE 1987

(millions of fiscal year 1975 dollars)

	FY 1975	FY 1976	Transition Quarter	FY 1977	FY 1978	FY 1979	Subtotal, 1975-79	Subtotal, 1980-2020	Total, 1975-2020
LMFBR									
Research and development									
FFTF	65	46	12	46	37	37	243	526	769
CRBR [a]	44	44	12	64	50	33	247	67	314
Support facilities	43	47	13	60	63	63	289	613	902
Engineering and technology									
Technology	52	53	15	57	60	63	299	588	887
Engineering	49	53	15	69	94	123	403	759	1,162
Cooperative projects									
CRBR [a]	21	73	5	155	160	140	554	200	754
NCBR [b]		17	5	5	18	64	87	189	276
Capital equipment	19	17	5	23	24	26	114	201	315
Construction projects									
FFTF	132	74					206		206
Plant component test facility		5		9	41	110	165	203	368
Radiation and repair engineering facility				9	18	14	41	5	46
Advanced fuel lab				9	18		27		27
Fuels and materials exam facility				23			23		23
Hot reprocessing pilot plant				2	7	28	37	239	276
Miscellaneous projects	15	18	3	20	28	17	101	91	192
Total LMFBR	440	430	80	551	618	718	2,836	3,681	6,517

Table 18 (continued)

	FY 1975	FY 1976	Transition Quarter	FY 1977	FY 1978	FY 1979	Subtotal, 1975–79	Subtotal, 1980–2020	Total, 1975–2020
SUPPORTING TECHNOLOGY									
Safety									
Research and development	37	40	12	46	52	58	245	646	891
Equipment	4	4	2	4	3	4	21	49	70
Construction									
Safety test facility		3		9	18	46	76	108	184
Transient reactor safety test facility				11	9	7	27		27
Advanced fuel technology	12	15	5	18	23	28	101	352	453
Total supporting technology	53	62	19	88	105	143	470	1,155	1,625
TOTAL LMFBR AND SUPPORT	493	492	99	639	723	861	3,306	4,836	8,142

a Clinch River Breeder Reactor.
b Near Commercial Breeder Reactor.
Source: USAEC, *Proposed Final Environmental Statement, WASH-1535*, vol. 4, p. 11.2-34.

Table 19

FORECAST AND REALIZED COSTS FOR THE
1958 TEN-YEAR PROGRAM OF CIVILIAN
NUCLEAR POWER DEVELOPMENT[a]

Reactor Type	Ratio of Realized to Forecast Expenditures on Construction, All Reactors of That Type Completed before 1967	Number of Reactors
Pressurized-water reactor	1.2	3
Boiling-water reactor	1.4	5
Heavy-water reactor	1.7	1
Gas-cooled reactor	1.9	1
Sodium graphite reactor	2.4	1
	>2.0	Canceled
Homogeneous reactor	1.6	1
	Unknown	Canceled
Organic-cooled reactor	2.1	1

a The estimates for the fast breeder reactor are not shown.
Source: MacAvoy, *Economic Strategy for Developing Breeder*, p. 114.

The AEC's Tradition of Cost Underestimation

These cost projections may not have fully accounted for future overruns if past experience with civilian nuclear power development is an accurate indicator. After examining various fiscal reviews of the civilian power program from annual publications of the AEC and from reports of the congressional Joint Committee on Atomic Energy, Paul W. MacAvoy presented the following summary on the forecast and realized costs for the 1958 ten-year program of civilian nuclear power R & D (see Table 19). All projects failed to stay within the estimated costs, and the average ratio of actual to design costs was greater than 1.5 for all projects.

The two projects that account for the largest share of the LMFBR program costs are the Fast Flux Test Facility (FFTF) and the CRBR. The construction cost of the FFTF, initially estimated at $87.5 million (1968 dollars) in 1968, rose to around $200 million (1973 dollars) [11] in 1973. In CB-74, it was estimated at $206 million (1974 dollars).

---

[11] Robert Gillette, "One Breeder for the Price of Two?" *Science*, vol. 182 (October 5, 1973), p. 38.

Similarly, the construction cost estimate of the CRBR has soared alarmingly:

> The first official estimate of its cost was about $400 million. In a 1972 Memorandum of Understanding its cost was estimated at $700 million, two thirds coming from the AEC. This estimate was $150 to $200 million higher than an AEC estimate only six months previous. In March 1974, it was reported that CRBR project officials are "focusing on some major steps that they hope will hold the total cost of the plant under $1.0 billion."[12]

In CB-74 the AEC estimated the CRBR's construction cost at $754 million and its total cost at $1.7 billion, both significantly higher than previous estimates.

In 1973, Merrill J. Whitman, chairman of the Federal Power Commission's advisory task force on energy conversion R & D and an AEC official, estimated that the construction of the near commercial breeder reactor would cost the government $200 million.[13] From Table 18, the government's share of the cost is estimated at $276 million. The capacity of the NCBR will be four to six times that of the CRBR and its construction cost will not be lower than that of the CRBR. It is very doubtful that the nuclear and utilities industries will absorb the major share of its construction cost.

Another $90 million,[14] which was not included in CB-74, is expected to be spent in direct assistance to utilities for their purchase of the first four commercial FBR power plants. Also excluded in the cost-benefit analysis of the LMFBR in CB-74 is the program cost for the gas-cooled fast reactor (GCFR) and the molten-salt breeder reactor (MSBR), which is estimated by the AEC to be $1.4 billion undiscounted. The AEC further stated that the technological activities for the GCFR and MSBR

> will continue into the early 1980's and will phase out approximately four years prior to the introduction of the LMFBR. It is assumed that by 1982–1983, there will be a high level of confidence that the LMFBR will be introduced and further work on alternatives and backups would not be necessary.[15]

---

12 Cochran and Tamplin, "NRDC Comments," in USAEC, *Proposed Final Environmental Statement*, WASH-1535, vol. 6, p. 38-174.

13 Merrill J. Whitman's estimate, quoted in Gillette, "One Breeder for Price of Two?", p. 38.

14 Gillette, "One Breeder for Price of Two?", p. 38.

15 USAEC, *Proposed Final Environmental Statement*, WASH-1535, vol. 4, p. 11.2-36.

A full-fledged program will cost $5.7 billion undiscounted. We conclude that the $1.4 billion cost for the GCFR and the MSBR programs, assuming they are conducted at the present low level and will be terminated by 1982–83, should be included in the costs of the LMFBR program, although our analysis does not explicitly reflect their benefits. With the fusion reactor probably to be introduced in the first decade of the next century, there is practically no justification for a parallel breeder program. In other words, the GCFR and MSBR programs being planned by the AEC only serve as a backup alternative and their costs are the insurance premium against the possibility of an unsuccessful LMFBR program. If the benefit of the LMFBR were $14.7 billion as computed by the AEC in CB-74 for the base case, a $1.4 billion insurance premium might be justifiable. However, if the LMFBR itself is not likely to produce any benefit, it is highly questionable that the GCFR and the MSBR should be conducted at such a low level. A careful economic and environmental analysis on these alternative breeders should be undertaken to determine whether their funding should be significantly reduced or increased. A sharp increase in support for the GCFR or the MSBR corresponds to a drastic reduction in LMFBR funding and the replacement of the LMFBR by the GCFR or the MSBR as the principal breeder design. By no means do we favor such a switch. But unless a detailed analysis is undertaken, the GCFR and the MSBR program should be further deemphasized. If the costs of the GCFR and MSBR are included, the benefits of the LMFBR shown at the bottom of Figure 7 would be further reduced by approximately $0.7 billion.

From the above consideration, it is apparent that the AEC traditionally underestimates the expected costs of projects. The commercial LMFBR will not be introduced until 1987 at the earliest. Based on the AEC's past record of cost projection and the assumptions employed in its projections, it is likely that the costs of the LMFBR program will be significantly higher than those projected in CB-74 thirteen years before the LMFBR's commercial introduction. The AEC stated: "Cost overruns and schedule slippages are, by their nature, impossible to predict accurately. Part of good program planning is to establish reasonable cost and schedule goals and to attempt to meet them." [16] It is usual for one to set cost and schedule goals and attempt to meet them. But a cost-benefit analysis should make realistic estimates based on past experience and future expectation. Allowing projected LMFBR program costs to soar from $2.96 billion in CB-68 to $3.59 bil-

---

[16] Ibid., p. 11.2-39.

lion in CB-70 and finally to $7.20 billion in CB-74 [17] and the LMFBR introduction date to slip from 1984 in CB-68 to 1986 in CB-70 and to 1987 in CB-74 reflects the unreliability of the AEC's projections.

Table 4 shows that AEC expenditures on the LMFBR, from its inception to the end of fiscal year 1974, total $1.64 billion (undiscounted and in current dollars). In a cost-benefit analysis, past or sunk cost is sunk forever. Nevertheless, it is still worthwhile to compare the total cost of a program with its benefit, especially for programs requiring a major commitment of national resources. It is possible to keep a program alive by continually underestimating its costs. Hindsight is infallible but often educational. It should be stressed that the present worth of the LMFBR's past costs now amounts to $3.3 billion, a figure which is calculated in the appendix.

Further, in the left block at the bottom of Figure 7, future LMFBR program costs are $4.7 billion, as estimated by the AEC in CB-74. All other costs discussed in this chapter are not included. In the right block, total program costs include past expenditures of $3.3 billion in addition to the future cost of $4.7 billion. Again, none of the cases shown in Figure 7 gives a benefit to the LMFBR.

---

[17] These costs are expressed in 1969 dollars. Supporting technology costs are allocated between LMFBR and other breeders proportionally. LMFBR expenditures for FY 1970 and 1971 are added to estimates in CB-70. Similarly, expenditures for FY 1970-74 are added to CB-74. Thus, all three estimates refer to LMFBR program costs starting in FY 1970.

6

CONCLUDING REMARKS

The Stauffer-Palmer-Wyckoff Report on Fast Breeder

Following completion of this manuscript, a study by Harvard economist T. R. Stauffer, R. S. Palmer of General Electric, and H. L. Wyckoff of Commonwealth Edison Company was released in which the authors estimated the net discounted benefit of the LMFBR to be $76 billion (1975 dollars),[1] even higher than the AEC's $14.7 billion (1974 dollars). A few comments on this study—called here the SPW study—and on those extremely high figures are in order.

The assumptions used by Stauffer, Palmer and Wyckoff are shown in Table 20. It is the use of a different uranium supply-versus-price schedule and different discount rates that accounts for most of the discrepancies in the results of the SPW study, the AEC's and ours.[2] The uranium supply shown in Table 20 is significantly lower than those shown in Table 7. The estimate of total resources at prices up to $65 per pound U_3O_8 is only 2.4 million tons, compared with the 7.5 million tons shown in Case B of Table 7 and the 4.2 million tons used by the AEC for the base case in CB-74. The argument against using such a low estimate of uranium resources is presented in Chapter 3 above. If the 4.2 million ton figure were used in the

---

[1] T. R. Stauffer, R. S. Palmer, and H. L. Wyckoff, study quoted in "New Study Packed with Good News for LMFBR," *Weekly Energy Report*, vol. 3, no. 10 (March 10, 1975), pp. 1-6. In an earlier circulated draft of the study, the figure given for the net discounted benefit of the LMFBR was $157 billion (see Stauffer, Palmer and Wyckoff, "An Assessment of the Economic Incentive for the Fast Breeder Reactor," draft, Harvard University, October 28, 1974).

[2] We also wonder, in the SPW study, how the net benefit for LMFBR can remain unchanged at $76 billion when the cost of uranium enrichment is cut in half (case 13 of SPW study). In CB-74, when the enrichment cost is reduced from $75/SWU (case 74) to $36/SWU (case 3), the gross benefit before R & D costs is reduced from $29.3 billion to $19.4 billion, a change of $9.9 billion.

Table 20

ASSUMPTIONS FOR THE BASE CASE OF STAUFFER, PALMER, WYCKOFF STUDY

Electrical growth rate: year 1975–2000		6%
year 2001–2020		4%
Inflation rate		6%
Discount rate: without inflation correction		6%
with inflation correction		12.36%
Advanced converter as a percent of nonbreeder capacity		25%

Uranium supply-versus-price schedule:

	Probable price range (1974 \$/pound U_3O_8)	
	\$10-15	\$15-65
Known reserves (kilotons)	275	425
Estimated additional reserves (kilotons)	450	1,250
Total (kilotons)	725	1,675

Coal price for new fossil capacity-base increase linearly (1975\$): year 1975	90¢/MBtu's
year 2020	\$1.35/MBtu's
Capital cost of coal plant (1975\$)	\$340/kWe
Capital cost of LWR plant (\$600/kWe in 1982)	\$385/kWe
Capital cost of LMFBR plant	\$1.25/LWR
Cost of new uranium enrichment capacity	\$70/SWU
Breeder plant efficiency	38%
Breeder introduction date	1989

Source: Uranium price from Stauffer et al., "Economic Incentive for Breeder," p. 2. The rest is from Stauffer et al., quoted in *Weekly Energy Report*, p. 6.

SPW study, the net discounted benefit for the LMFBR would be in agreement with the AEC's figure, using a discount rate of 6 percent.

The discount rate used in the SPW study, based on social rate of time preference (SRTP), is 6 percent and is explained as follows:

> For example, if the members of society will put their money in the bank at an interest rate of 4% (uninflated), this is a measure of the value they put on deferring immediate consumption. Hence, the discount rate defined by the SRTP is closely related to the cost of debt money.[3]

---

[3] Stauffer et al., "Economic Incentive for Breeder," p. 16.

But, should the rate be 6 percent? Is the interest rate paid by banks a good measure of the SRTP? The cash people keep in banks is idle and surplus cash, which can be withdrawn with little penalty at any time, whereas, for the LMFBR program, money has to be invested for the long term and continuously. Any premature withdrawal will not only forfeit the interest but also the original investment. Thus, the bank interest rate is not a good measure of the SRTP.

What prevents us from taking the other extreme, namely, the interest rate of consumer loans? For example, one can say that the funds for the LMFBR program come from taxation. Because of mandatory support of the program through taxes, some people have to borrow money for their immediate consumption at an interest rate of 18 percent. The SRTP should, therefore, be at least 18 percent! The above argument is just as valid or invalid as the one argument for using the bank interest rate. The SRTP should be a weighted measure of the future benefits that society considers indifferent to its immediate consumption. Such a weighted measure will lead to a discount rate of well above 6 percent. As long as other governmental projects are using a 10 percent discount rate after inflation and before tax, the LMFBR program should be evaluated on the same basis for equitable comparison, unless the LMFBR draws a significant portion of resources from idle and low-return resources, which is not the case. It is not beneficial to society to support a project that can earn only a 6 percent return while these resources displaced by the project, if left alone, can earn a substantially higher rate, as discussed in Chapter 2.

Findings

The present study raises serious doubts about the assumptions and projections employed in the AEC's cost-benefit analyses on the Liquid Metal Fast Breeder Reactor Program. Many of the AEC's projections lead to an unrealistically large benefit from the LMFBR: Uranium resources are underestimated. The high-temperature gas reactor is artificially restricted to a low level of participation in the future electric power system. Future energy demand is overestimated. The plant capital cost of the LMFBR is decreased too rapidly to fit any reasonable learning curve. And finally, schedule slippages and cost overruns are not adequately reflected in the analyses.

It thus appears that the LMFBR's high efficiency in uranium utilization is not sufficient to compensate for its higher plant capital and program costs. The LMFBR program yields no net discounted economic benefits. Moreover, according to Cochran, the environmental benefits claimed by the AEC also do not exist.[4] In view of the above considerations, can one justify support for the LMFBR program at the level proposed by the Atomic Energy Commission? Recommended expenditures for this program over the next five years total $2.6 billion (in undiscounted current dollars), equal to 63 percent of the nuclear fission R & D budget, 26 percent of the energy budget, and almost twice the nuclear fusion reactor budget. This is the highest expenditure among all the federal energy R & D programs.

Should we continue to treat the fast breeder reactor as our top priority program, knowing that it will not supply electricity until 1987 at the earliest, that it will not help alleviate our current energy crisis and that, once introduced, it will probably be displaced or replaced by fusion reactors in ten or twenty years? It is true that $5 billion or even $10 billion (discounted at 10 percent to mid-1974) for the development of the breeder reactor is only a small percentage of the national electrical energy cost of $200 billion (discounted at 10 percent to mid-1974) or $6,000 billion (undiscounted) in the period from 1974 to 2020. The amount looks even smaller when compared with $140 billion (undiscounted) in military aid and $350 billion (undiscounted) in total expenditures spent over the years in Vietnam.

We are convinced that national energy programs deserve much stronger support than they currently receive because energy is an essential commodity which dictates the future of our economy and our livelihood. If funds for energy programs were abundant and if all other energy programs were adequately funded, the fast breeder program could be supported on the basis of the same philosophy that underlies support for basic research and development. Unfortunately, all of our energy programs have to compete with each other within the framework of present and future federal energy budgets. Supporting the LMFBR program at the currently planned level will require reducing our commitment to other worthwhile energy programs. The eventual loss to society will not be the $5 billion or $10 billion that the LMFBR program will cost. Rather, it will be the difference between the national energy cost with the LMFBR included in the electrical energy system and that of an alternative energy system which comes about from the release of funds from the LMFBR program. The difference may be many times $10 billion. The concern, therefore, is not so much the cost of the LMFBR program but rather the optimal mix of programs, under a given energy budget, that will provide us with adequate energy at the lowest cost.

---

[4] Cochran, *Fast Breeder Reactor*, pp. 223-29.

Should a portion of the funds for the FBR program be transferred to other programs, such as the safety of the LWR, the pollution abatement of coal-fired power plants, the in-depth assessment of uranium resources, the improvement of the HTGR, the accelerated development of fusion reactors, the increase in domestic production of oil and gas, the massive substitution of coal for oil and gas, the exploitation of renewable energy sources, and the conservation of energy and energy resources? In order to fulfill the nation's electric energy requirement, should we reduce the FBR program to a low level, regard it as a backup option, and concentrate efforts and resources on improvements in the LWR and coal-fired plants for the short term, on the HTGR for the intermediate term, and on the development of the fusion reactor for the long term? The answer to both questions is yes. Before the nation commits itself more heavily to the FBR, it can afford to wait another five to ten years for better projections of future energy demand, better estimates of uranium resources, and a clearer determination of the feasibility of an economically and environmentally acceptable commercial fusion reactor. However, further studies have to be conducted and alternatives examined in a cautious manner before one can spell out in detail the energy budget reallocation and develop an optimal national energy policy upon which the welfare of this generation and future generations so vitally depends.

APPENDIX
Present Worth of the LMFBR's Past Costs

In order to calculate the present worth of the LMFBR's past costs, one must express all the expenditures in terms of constant dollars, which is mid-1974 dollars in the present case, and employ appropriate discount rates for the previous years. If the comptroller general's guidelines and assumptions (discussed in Chapter 2) are followed, and if the interest rate for corporate bonds is assumed to be 1.5 percentage points above that of long-term government bonds, the following formula closely approximates the relationship between the average interest rate of long-term government bonds, G, and the discount rate, D, for a particular year:

$$D\% = 5\% + 0.75G\%$$

An accurate formula, if it exists at all, is not necessary for the rough estimation presented here. Table A-1 summarizes our computation. Past expenditures are totaled at $3.3 billion.

Table A-1

PRESENT WORTH OF THE LMFBR'S PAST COSTS, 1947–74

($ in millions)

Fiscal Year	LMFBR Past Costs (in millions) Current dollars	LMFBR Past Costs (in millions) Mid-1974 dollars	Average Long-Term Government Bond Yield (in percent)	Discount Rate (in percent)	Cumulative Present Worth of Past Costs (mid-1974 dollars)
1947	.9	2.0	2.3	6.7	2.0
1948	.6	1.3	2.5	6.9	3.4
1949	1.1	2.3	2.3	6.7	5.9
1950	2.4	4.9	2.3	6.7	11.3
1951	.7	1.4	2.6	6.9	13.4
1952	3.5	6.7	2.6	7.0	21.0
1953	3.4	6.4	2.9	7.2	28.8
1954	3.9	7.3	2.6	7.0	38.1
1955	5.0	9.1	2.8	7.1	49.9
1956	5.8	10.1	3.1	7.3	63.5
1957	7.6	12.9	3.5	7.6	81.0
1958	16.1	27.0	3.4	7.6	114.2
1959	21.8	35.8	4.1	8.1	158.6
1960	20.3	32.7	4.0	8.0	204.1
1961	14.3	22.8	3.9	7.9	243.5
1962	17.1	27.0	4.0	8.0	289.8
1963	26.0	40.3	4.0	8.0	353.2
1964	26.7	40.6	4.2	8.1	422.1
1965	34.1	51.0	4.2	8.2	507.4
1966	42.6	62.0	4.6	8.5	610.8
1967	78.0	110.7	5.0	8.8	773.1
1968	95.0	128.2	5.3	8.9	969.1
1969	106.0	133.4	6.0	9.5	1,189.2
1970	120.0	144.5	6.6	10.0	1,446.7
1971	176.0	208.4	5.8	9.3	1,799.1
1972	205.0	236.0	5.6	9.2	2,202.8
1973	253.8	278.7	6.0	9.5	2,684.3
1974	356.8	356.8	6.8	10.1	3,296.3

Source: LMFBR past costs are from Table 1 and bond yields are from Moody's *Municipal and Government Manual,* vol. 2 (1974), p. a16.